QUANTITATIVE BIOSCIENCES COMPANION IN PYTHON

Dynamics across Cells, Organisms, and Populations

COMPANION IN PYTHON

QUANTITATIVE BIOSCIENCES

Dynamics across Cells, Organisms, and Populations

JOSHUA S. WEITZ, NOLAN ENGLISH,
ALEXANDER B. LEE, AND ALI ZAMANI

Princeton University Press
Princeton and Oxford

Published by Princeton University Press
41 William Street, Princeton, New Jersey 08540
99 Banbury Road, Oxford OX2 6JX

press.princeton.edu

ISBN (pbk.) 9780691255675
ISBN (e-book) 9780691259611

Library of Congress Control Number: 2023946866

British Library Cataloging-in-Publication Data is available

Editorial: Sydney Carroll and Johannah Walkowicz
Production Editorial: Terri O'Prey
Text Design: Wanda España
Cover Design: Wanda España
Production: Jacqueline Poirier
Copyeditor: Jennifer McClain

Cover image: Simon Dack News / Alamy Stock Photo

This book has been composed in MinionPro and Omnes

10 9 8 7 6 5 4 3 2 1

CONTENTS

Preface ix
The goal ix
You can do it xi
Acknowledgments xiii

I MOLECULAR AND CELLULAR BIOSCIENCES 1

1 Fluctuations and the Nature of Mutations 3

1.1 Hands-on approach to mutations and selection 3
1.2 Sampling from provided distributions 4
1.3 Sampling from custom distributions 6
1.4 Comparing binomial and Poisson distributions 8
1.5 The start of dynamics 10
1.6 Inferring parameters from data 11
Solutions to Challenge Problems 15

2 Bistability of Genetic Circuits 21

2.1 Continuous models of cellular dynamics and gene regulation 21
2.2 Simulating coupled ordinary differential equations 23
2.3 Qualitative analysis of nonlinear dynamical systems 26
2.4 Evaluating the local stability of equilibria 29
2.5 Bistability and bifurcation diagrams 32
Solutions to Challenge Problems 36

3 Stochastic Gene Expression and Cellular Variability 41

3.1 Simulating stochastic gene expression 41
3.2 Poisson processes: Finding the time of the next event 42
3.3 A theory of timing given multiple stochastic processes 45
3.4 Gillespie algorithm applied to a gene expression model 49
3.5 Loading and saving data 56
Solutions to Challenge Problems 56

4 Evolutionary Dynamics: Mutations, Selection, and Diversity 61

4.1 Modeling evolutionary dynamics 61
4.2 Transition matrices in Markov processes 62
4.3 The Wright-Fisher model 69
Solutions to Challenge Problems 76

II ORGANISMAL BEHAVIOR AND PHYSIOLOGY 81

5 Robust Sensing and Chemotaxis 83

5.1 Toward chemotaxis in single-celled organisms 83
5.2 Enzyme kinetics 84
5.3 Time-dependent functions in differential equations 87
5.4 Probability distribution redux 88
5.5 *E. coli* movement 93
Solutions to Challenge Problems 99

6 Nonlinear Dynamics and Signal Processing in Neurons 105

6.1 Computational neuroscience 105
6.2 The Hodgkin-Huxley model 107
6.3 Firing without a current 112
6.4 Neuron dynamics: Thresholds in magnitude and time 114
6.5 Technical appendix 116
Solutions to Challenge Problems 119

7 Excitations and Signaling, from Cells to Tissue 125

7.1 Excitable media: From localized to spatial dynamics 125
7.2 FitzHugh-Nagumo: The ODE model 126
7.3 FitzHugh-Nagumo: One-dimensional PDEs 131
Solutions to Challenge Problems 139

8 Organismal Locomotion through Water, Air, and Earth 143

8.1 Introduction 143
8.2 The internal origins of movement 144
8.3 Orbits in configuration space 146
8.4 From Borelli to Newton and back again 147

8.5 The greatest gait of all 154
Solutions to Challenge Problems 154

III POPULATIONS AND ECOLOGICAL COMMUNITIES 159

9 Flocking and Collective Behavior: When Many Become One 161

9.1 Agent-based models and emergence in flocks 161
9.2 The Vicsek model 163
9.3 Flocking dynamics 165
9.4 Bonus: The power of leadership 168
Solutions to Challenge Problems 169

10 Conflict and Cooperation Among Individuals and Populations 175

10.1 Strategies, games, and populations 175
10.2 Mean field replicator dynamics of microbial games 178
10.3 Stochastic versions of microbial games 180
10.4 Type VI secretion—a killer game, in space 185
Solutions to Challenge Problems 191

11 Eco-evolutionary Dynamics 195

11.1 From predation events to population dynamics 195
11.2 Ecological dynamics when evolution is fast 196
11.3 Functional responses—a microscopic approach 200
Solutions to Challenge Problems 204

12 Outbreak Dynamics: From Prediction to Control 211

12.1 Outbreaks: From deterministic models to stochastic realizations 211
12.2 Epidemic modeling—fundamentals 212
12.3 Stochastic epidemics 218
Solutions to Challenge Problems 221

IV THE FUTURE OF ECOSYSTEMS 227

13 Ecosystems: Chaos, Tipping Points, and Catastrophes 229

13.1 Modeling complexity: An enabling view 229
13.2 Small differences, big effects 230

13.3 Explosive growth and population catastrophes 236
13.4 Small models of a big climate 240
13.5 Coda 243
Solutions to Challenge Problems 243

Bibliography 251

PREFACE

THE GOAL

This computational laboratory guide accompanies the textbook *Quantitative Biosciences: Dynamics across Cells, Organisms, and Populations*. The guide is written with students and early career scientists in mind. The near-term goal is simple: to translate biological principles and mathematical concepts into computational models of living systems. The use of computation is key. Developing computational models both democratizes and broadens the range of students from diverse backgrounds who can meaningfully integrate mathematical principles and biological concepts. In that sense, the long-term goal of this guide is to change the culture of how biology is taught and how biological research is conducted in practice.

As developed, the course upon which both the textbook and these lab notes are based includes a recurring structure. Each week is centered upon a focal scale and biological question. Are mutations dependent on selection or independent of selection? How do bacteria sense and respond to their environment? How do neurons and cardiac cells filter and integrate signals? How do individuals in a collective spontaneously move in flocks? How does rapid evolutionary change modify the dynamics of populations? These and other questions motivate a series of two lectures totaling approximately three hours that include a mix of biologically focused slides, an introduction to and derivation of mathematical concepts, and a journal club–like paper discussion. Then, on the third day, the class meets in a "laboratory" format for another three hours. Each student has their own computer. In front of them is a student version of that week's laboratory guide. There are no files to download, at least not typically. Instead, the laboratory guide includes code in MATLAB, Python, or R that engages with the themes and questions of the week. The students enter the code because it turns out that the act of typing reinforces concepts they already understand and highlights concepts or practices they do not yet understand.

For example, in the first week, we explore evidence that reveals whether mutations are dependent on or independent of selection. The textbook details both the biological evidence and mathematical concepts that underlie Luria and Delbrück's (LD) conclusion that mutations are random and independent of selection (Luria and Delbrück 1943). Yet, if students are to truly commit these concepts into their own practice, they must strive to re-create them. In a Quantitative Biosciences class, such an effort could involve asking students to (re)derive the Luria and Delbrück distribution or its moments. However, doing so would likely preclude many students who grasp the core idea underlying the LD mechanism but find rigorous mathematical derivations do not add to their intuition. Instead, we take a different tack. We ask students to translate the mathematical concepts into a computational model, and then probe the qualitative and quantitative behavior of the model in regimes that reflect the biological system at hand. The art of building the model reinforces and deepens student intuition. This is not just a matter of convenience. For many systems of interest, e.g., neuronal firing, flocking of organisms, or eco-evolutionary dynamics, there may not exist a closed form solution to derive (or if one exists, it may reflect only a partial, asymptotic, or even approximate solution). In essence, computation is an imperative, not just an alternative.

This book embraces the imperative to create models of living systems for yet another purpose: to bridge the gap between receiving information from the instructor and reaching a deeper understanding. This bridging of the gap has been described by the eighteenth-century chemist Joseph Priestly as "something that cannot be described in words." David Kaiser (2005) elaborated on this concept in assessing the history of the dissemination of Feynman diagrams in the post-WWII era:

> Experimentalists must work hard to hone something like artisanal knowledge or craft skill in addition to an understanding of general principles. Historians and sociologists have argued that tacit knowledge plays a central role when it comes to replicating someone else's instruments, even when the would-be replicator is already an expert experimentalist or instrument maker.

Replace the word *experimentalists* with *quantitative bioscientists* and the word *instruments* with *code* and this quote animates the core of this hands-on computational laboratory guide. The tacit knowledge helps steer students toward good coding practice, clear and understandable modeling, and accessible visualizations of model results.

And yet there is one more objective of this computational guide. As taught in a course format, the computational laboratory guide is meant to help students prepare for homework that encourages independent thinking, problem solving, and ultimately independent research. To do so requires that students are prepared with a diverse repertoire of computational skill sets. This rationale is similar to that underlying the proliferation of tool-centric coding workshops in biology. As is apparent, if students cannot load, analyze, and visualize their data in a rigorous and repeatable manner, that will undermine the quality of well-designed experiments and sampling schemes. Yet analyzing data in the absence of biological questions can become a hollow enterprise. If students are to understand how feedbacks in living systems lead to emergent phenomena not necessarily embedded in the properties of individual components, then they need to develop a diverse repertoire of simulation approaches, to build the appropriate simulation model at the appropriate scale.

The goal of teaching practical skills as a means to increase tacit knowledge is embedded in each module. The modules themselves are not organized by method but by problem, in parallel to the organization of the main text. But the methods do in fact build upon each other. Each module introduces and/or reinforces different practical approaches to develop computational models of living systems. Students who work their way through this book should expect to gain practical expertise in the following methods:

- Sampling from probability distributions
- Stochastic branching processes
- Continuous time modeling
- Local stability analysis for nonlinear dynamical systems

- Stochastic modeling via the Gillespie algorithm
- Markov chains
- Bifurcation analysis
- Excitable system dynamics
- Partial differential equations
- Comparing stochastic to continuous models
- Agent-based simulations
- Discrete time dynamics—including the emergence of chaos

This is a non-trivial list. The mathematics underlying each approach could involve an entire course. But that is not the point. The point is to help students integrate each of these concepts when necessary into usable and principled code. In essence, the start of the title—*Quantitative Biosciences*—is meant to tell the story, emphasizing the essential integration of mathematical reasoning and computation in enabling understanding and eventually the ability of students to make advances of their own. This is a middle path, but one that we hope proves productive in the long term. Students interested in advanced studies in stochastic processes, nonlinear dynamics, or time series analysis are encouraged to pursue them. Perhaps they might be even more likely to do so after taking this course.

YOU CAN DO IT

After some debate and sufficient reflection, we have decided to replicate the course experience by organizing each chapter in the format of a student version. This means that chapters are structured as guides with questions, rather than manuals with solutions. Typically, the instructor version with solutions to challenge problems is handed out at the end of class. Here the answers are provided at the end of chapters. To get the most out of this book, we recommend that you try, as long as is possible, to avoid turning to these pages. Think of it as your own marshmallow test. The answers are there, but they are not the only answers. More than anything, the answers are there to ensure that when you get stuck there is a light to help keep you going along the path.

What then should you, the reader, do? Our recommendation: Just begin. This laboratory guide comes in three flavors: MATLAB, Python, and R. The descriptions and mathematics are common to all, so it is up to you to choose a language and stick with it. As applicable, source code is available at the book's website, free for download, use, and reuse. In addition, an optional tutorial in each language is also available on the website, and may represent a necessary primer for some students before beginning. Many other tutorials exist—but a lack of deep experience in coding should not prevent you from beginning. Why? Because you can do it.

ACKNOWLEDGMENTS

The central goal of this computational guide is to help students develop the tacit knowledge needed to explore living systems across scales. As noted in the accompanying textbook, this guide would not have been possible without the input of colleagues, students in the Quantitative Biosciences (QBioS) PhD program at Georgia Tech, and those who utilized material under development as part of undergraduate and graduate programs and as part of short courses in the United States, France, Italy, and Brazil. Special thanks go to QBioS and Georgia Tech students who provided critical feedback that has shaped the need for the book and the nature of the book itself: Qi An, Akash Arani, Emma Bingham, Pablo Bravo, Alfie Brownless, Rachel Calder, Alexandra Carruthers-Ferrero, Hyoann Choi, Ashley Coenen, Shlomi Cohen, Raymond Copeland, Sayantan Datta, Kelimar Diaz, Robert Edmiston, Shuheng Gan, Namyi Ha, Hayley Hassler, Maryam Hejri Bidgoli, Lynn (Haitian) Jin, Elma Kajtaz, Cedric Kamalseon, Katalina Kimball-Linares, Tucker J. Lancaster, Daniel A. Lauer, Zewei Lei, Guanlin Li, Hong Seo Lim, Ellen Liu, Lijiang Long, Katie MacGillivray, Jiyeon Maeng, Andreea Magalie, Pedro Márquez-Zacarías, Zachary Mobille, Daniel Muratore, Carlos Perez-Ruiz, Aaron R. Pfennig, Rozenn Pineau, Brandon Pratt, Joy Putney, Aradhya Rajanala, Athulya Ram, Elisa Rheaume, Rogelio A. Rodriguez-Gonzalez, Benjamin Seleb, Varun Sharma, Benjamin Shipley, Cassie Shriver, Michael Southard, Sarah Sundius, Disheng Tang, Stephen Thomas, Kai Tong, Akash Vardhan, Hector Augusto Velasco-Perez, Ethan Wold, Fiona Wood, Leo Wood, Siya Xie, Christopher Zhang, Mengshi Zhang, Conan Y. Zhao, and Baxi Zhong.

Thank you to Van Savage, David Murrugarra, Rafael Peña Miller, and Carles Tardío Pi for their comprehensive review of the textbook and laboratory guides. Van gets a double thanks for his willingness to try out this material in formative stages—thanks also to Tianyun Lin for facilitating feedback from UCLA students that has been essential to refining the format and content of the book. Thank you to colleagues at Georgia Tech, especially those in the Physics of Living Systems program, for their input that shaped the computational lab guide material over many years, especially Flavio Fenton, J. C. Gumbart, Simon Sponberg, and Daniel Goldman, as well as to current and former colleagues in Biological Sciences, especially Will Ratcliff as well as multiple group members who helped teach part of the course and whose input was critical to improve the material: Stephen Beckett, David Demory, Adriana Lucia Sanz, Jeremy Harris, Joey Leung, and Jacopo Marchi. We have tried to follow the good counsel of our colleagues—any remaining mistakes are ours alone.

The time and resources to develop this computational lab guide have been made possible, in part, by grants and support from the National Science Foundation (NSF) Physics of Living Systems, Biological Oceanography Dimensions of Biodiversity, Bridging Ecology and Evolution programs, NIH NIGMS, NIH NIAID, Army Research Office, Charities in Aid Foundation, Marier Cunningham Foundation, Chaire Blaise Pascal Program of the Île-de-France, Mathworks Corporation, Simons Foundation, Centers for Disease Control and Prevention, A. James & Alice B. Clark Foundation, and the Burroughs Wellcome Fund.

Finally, thank you to the entire staff at Princeton University Press, including Alison Kalett, Sydney Carroll, and the copyeditors, illustrators, indexers, and production specialists who have elevated this material into an integrated whole.

Part I

Molecular and Cellular Biosciences

Chapter One

Fluctuations and the Nature of Mutations

1.1 HANDS-ON APPROACH TO MUTATIONS AND SELECTION

The goal of this lab is to simulate a growing bacterial population, including the ancestral "wild type" as well as mutants generated de novo during the growth process. The core techniques are straightforward: connecting the simplest model of exponential growth with stochastic events. To do so requires a few techniques, all centered on the ramifications of sampling from random distributions using Python. As you will see, learning how to sample from random distributions will be relevant in many biological systems. Indeed, being able to simulate stochastic dynamics is key for simulating biological systems at scales from molecules to organisms to ecosystems. Hence, this opening chapter introduces basic concepts that are used throughout the laboratory guide. This chapter also serves another function: to link the material in the textbook with the homework.

The laboratory will prepare you to build components of two categories of mutational models, as illustrated in a generalized schematic form in Figure 1.1. These initial components form the basis for the homework problems presented in the main text. In this figure, the left panel illustrates a branching process in which an individual bacterium in generation $g = 0$ divides so that there are two bacteria in generation $g = 1$, four bacteria in generation $g = 2$, and so on such that there are 2^g bacteria after g generations. Of these, a fraction of the offspring may be different than the ancestral wild type. These different bacteria are referred to as *mutants*. Notably, in this model, mutants give rise to mutant daughter cells and not to wild-type cells. The right panel illustrates an alternative model of mutation, in which many bacteria in a single generation undergo some stochastic change, i.e., a mutation, rendering a small number of bacteria into mutants. This latter case may be related to a phenotypic change, e.g., exposure to a virus or chemical agent. How to build models of both kinds, how to compare them, and how to reconcile the predictions of such models with experimental data from Luria and Delbrück form the core of this laboratory.

The key aim of this laboratory is to begin a process to relate the mechanism by which mutants are generated with signatures that can be measured. These signatures may include the mean as well as the variance in the number of mutants between parallel experiments.

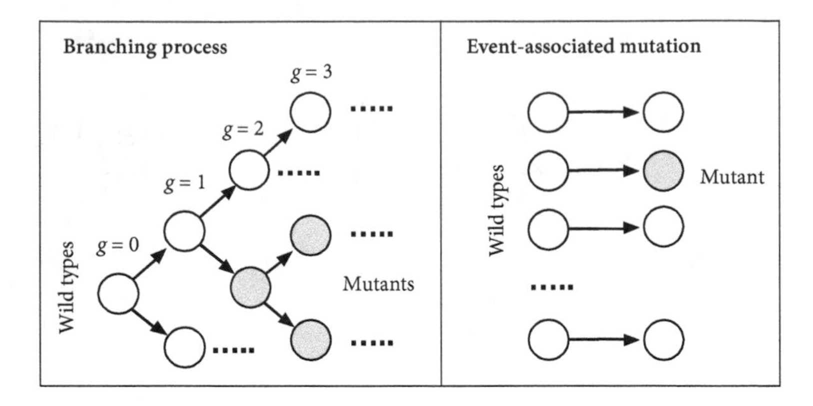

Figure 1.1: Stochastic models of mutation: Mutations are independent of selection (left) or dependent on selection (right). (Left) Branching process in which a single (or small) number of wild-type bacteria (empty cells) divide and occasionally mutate; the mutants (shaded) also divide. (Right) Mutation occurs randomly among a large population given interaction with a selective pressure, leading to a small fraction of mutants.

1.2 SAMPLING FROM PROVIDED DISTRIBUTIONS

In order to simulate stochastic processes, such as mutation in a population, one must repeatedly sample random numbers. Random numbers can be generated by any modern programming language. In doing so, it is possible to use built-in functions or to manipulate the generated random numbers to ensure they have a specified mean, variance, and higher-order moments. For example, to randomly sample a number between 0 and 1, use the command

```
import numpy as np
import matplotlib.pyplot as plt

np.random.rand()
```

Do this a few times. Each number is different. But generating multiple random numbers one at a time is unnecessary. Instead, generating multiple random numbers can be done automatically; e.g., use the following commands to randomly sample 100 points between 0 and 1:

```
randvec = np.random.rand(1,100)
```

or

```
randvec = np.random.rand(100,1)
```

These commands will generate a set of 100 random numbers in either a row or a column.

It is also possible, as shown in the introductory coding demos available on the book's website, to generate random matrices. Use the following command to generate a random matrix of size $m \times n$ array:

```
randarray = np.random.rand(m,n)
```

As is apparent, the shape of the matrix can be specified in terms of the number of rows `m` and columns `n`. If the code does not work, that is probably because you have not yet defined the size; do so a few times and see how easy it is to generate distinct random matrices. Note for future reference that two arrays must be the same size in order to perform element-wise operations (e.g., addition, subtraction, or element-by-element multiplication). Also note the names of variables—they tend to be descriptive. This is a good practice because it makes code easier to read, modify, and reuse. The first challenge problem should help get you more comfortable working with the core features of random distributions.

CHALLENGE PROBLEM: Properties of Random Distributions

What is the mean value of a single instance of invoking `np.random.rand`? Similarly, what is the variance? Once you have identified the mean and variance, plot the distribution of numbers generated by `np.random.rand` by sampling a large number of points (10^4) and then using the `plt.hist` function to generate a histogram. What shape is the distribution? How does it change as you change the number of bins for the histogram?

Python also allows sampling different distributions than the uniform distribution. As one exercise, plot the distribution of the output for the following functions: (i) standard normal distribution with a mean of 20 and standard deviation of 5 using

```
np.random.normal
```

and (ii) the Poisson distribution with rate parameter $\lambda = 20$ using

```
np.random.poisson
```

Examples of the outputs can be seen in Figure 1.2.

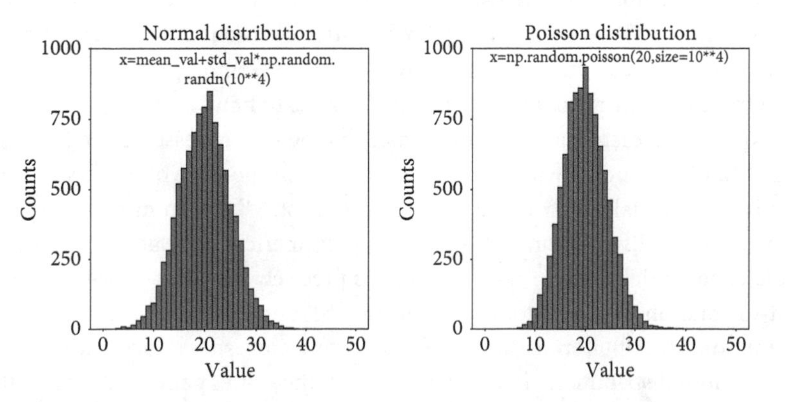

Figure 1.2: Sampling from random distributions, including the normal distribution (left) and the Poisson distribution (right).

It is possible to shift the range of randomly generated numbers using relatively simple operations, generating arbitrary variations (in range and location) of preexisting distributions. This may be useful in many circumstances, not only within the context of the LD problem. The following challenge problem provides an opportunity to build your intuition for manipulating and generating random numbers with distinct means and ranges.

CHALLENGE PROBLEM: Random Number Generation

This problem focuses on modifying the means and ranges of random numbers by modulating the output of built-in random number functions.

- Generate 1000 random numbers equally spaced between 0 and 5.
- Generate 1000 random numbers equally spaced between 2 and 7.
- Generate 1000 random numbers equally spaced between −5 and 5.

In each of these cases, use the built-in random number generator and then simple arithmetic (i.e., addition, subtraction, and multiplication) to transform the random numbers to specified ranges. You can do it!

1.3 SAMPLING FROM CUSTOM DISTRIBUTIONS

Python offers the option to generate specialized distributions. However, it is also possible to sample from "custom" distributions, i.e., both parametric and non-parametric distributions. One way to do so is to leverage the *cumulative distribution function*, or cdf. The cdf at a point, x, gives the probability of observing a value less than or equal to x. Formally, if $p(x)\mathrm{d}x$ is the probability of observing the random variable between x and $x + \mathrm{d}x$, then the cdf is

$$P(x) = \int_{-\infty}^{x} p(y)\mathrm{d}y \tag{1.1}$$

where y is a "dummy" variable used here for notational purposes of integrating over the probability distribution. The cdf is a monotonically increasing function with a range between 0 and 1. These constraints allow random sampling from arbitrary distributions if one is provided with the cdf in advance, by leveraging properties of the uniform distribution. An ideal way to illustrate this is via the exponential distribution.

The exponential distribution arises in many biological processes. For example, for processes that randomly occur with a constant rate λ, then the time of the first occurrence of an event is exponentially distributed such that $p(x) = \lambda e^{-\lambda x}$, given mean time $1/\lambda$. The cdf of the exponential distribution is $1 - e^{-\lambda x}$. Most numerical software tools have packages to sample exponential random numbers; this is precisely why it is instructive to compare the built-in solution to the custom solution. Indeed, one can think of the cdf of the exponential random distribution as having a one-to-one correspondence with the cdf of the uniform random distribution. That is, whereas half the values generated by a uniform random distribution will be <0.5, that is not true for an exponential distribution. Instead, given the shape parameter λ, then half the values of an exponential distribution will have values $x < x_u$ such that $1 - e^{-\lambda x_u} = 0.5$. This insight can help move from one distribution to the other.

To sample random numbers from the exponential distribution, first sample from the uniform distribution between 0 and 1.

```
probsamp = np.random.rand()
```

Think of this as a random value of P, which we denote as c_u. By randomly sampling the cdf of the uniform random distribution, the next question becomes: what value of the exponentially distributed random variable x_e corresponds to that point in the cdf? To answer this question requires that we invert the cdf, i.e., $P = 1 - e^{-\lambda x_e}$, to obtain an equation of x_e in terms of the cdf. To show this in action, denote c_u as the randomly selected value from the cdf of the uniform distribution. To map the cdf of the uniform distribution onto the cdf of the exponential distribution (our custom distribution) requires that $c_u = 1 - e^{-\lambda x_e}$. For x_e to be an exponentially distributed random number requires transforming the uniform random numbers into the exponentially distributed random numbers we would like to generate:

$$
\begin{aligned}
1 - e^{-\lambda x_e} &= c_u \\
e^{-\lambda x_e} &= 1 - c_u \\
-\lambda x_e &= \log 1 - c_u \\
x_e &= \frac{-\log 1 - c_u}{\lambda}
\end{aligned}
\tag{1.2}
$$

This gives $x_e = -\frac{1}{\lambda}\log(1 - c_u)$. This converts the sampling distribution of `rand` (i.e., c_u) to an exponential distribution. In order to use this repeatedly, it will be convenient to make and save a function:

```
def rand2exp(probsamp,my_lambda):
    # Do not overwrite the lambda Python function
    # lambda is the rate of the Markov process
    # probsamp are uniformly random sampled numbers
    return -1/my_lambda*np.log(1-probsamp)
```

The following section leverages the prior code snippet for using uniform sampling to generate exponentially distributed numbers given a process with rate $\lambda = 1/10$:

```
my_lambda=1/10
probsamp=np.random.rand(10**3)
expprobsamp = rand2exp(probsamp,my_lambda)
```

Now it is time to see if any of this works—via a challenge problem.

CHALLENGE PROBLEM: Comparing Exponential Random Sampling

Compare the distribution of 10^3 exponentially distributed random numbers using the cdf-based method to the distribution using the following built-in Python command:

```
my_lambda=1/10
pythonexprnd=np.random.exponential(1/my_lambda,10**3)
```

Note that the function `random` allows sampling from a number of different common distributions. (Hint: Another helpful function is the *empirical cumulative distribution function*, or `ECDF`, from `statsmodels.distributions.empirical_distribution`. This is useful in generating cumulative distributions.) If your code is working, it should look like the following:

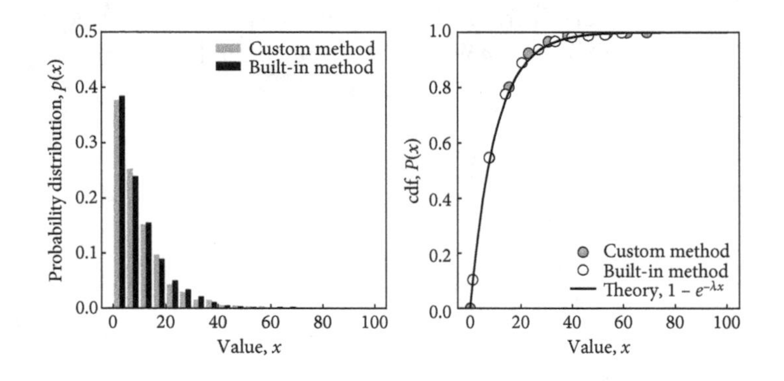

These figures show a comparison of customized sampling and built-in exponential random sampling via probability distributions (left) and cumulative distributions (right). For the cdf, the expected distribution is shown as a solid black line.

1.4 COMPARING BINOMIAL AND POISSON DISTRIBUTIONS

Binomial distributions result from counting the number of occurrences given independent samples with probability of occurrence p. For example, consider a mutation probability of $p = 10^{-8}$. Irrespective of whether mutations are independent of or dependent on selection, it would typically take a large number of cell divisions (or cells) for a mutant to appear. Using a binomial distribution, one could, in theory, predict the number of mutants expected to occur in a single round of cell division or given an exposure of a large collection of n cells to a selective force. Formally, the binomial distribution denotes the probability that k events occur out of n trials given the per trial probability p. This distribution is

$$P(k) = \binom{n}{k} p^k (1-p)^{n-k} \tag{1.3}$$

where $\binom{n}{k}$ denotes the number of unique ways of choosing k of n elements (i.e., the binomial coefficient). However, if occurrences are rare and the number of samples, n, is large, then the binomial distribution converges to the Poisson distribution with shape parameter $\lambda = np$ (this shape is the expected mean number of events in n trials). To see this computationally, compare the cdfs of the sample of repeated binomial sampling to repeated Poisson sampling with varying n. For example, use the following code to obtain and plot the cdf for binomial random numbers given 100 trials each with probability $p = 0.2$:

```
n=100
my_lambda = 20
p=my_lambda/n
numsamps=10**3
binosamps = np.random.binomial(n,p,numsamps)
sortbino = np.sort(binosamps)
cdfbino = np.arange(1,numsamps+1)/numsamps
plt.plot(sortbino,cdfbino,color='k',linewidth=2)
```

Here `binomial` samples binomial random numbers, and sorting the result allows for an explicit calculation of the empirical cdf (without using a built-in function).

CHALLENGE PROBLEM: Comparing the Binomial to the Poisson

Compare the binomial and Poisson cdfs for $n = 40$, 100, and 1000 in each case, assuming there is an expected number of 20 events such that the probability per event decreases from 0.5 to 0.2 to 0.02, respectively. If your code is working, it should look something like this:

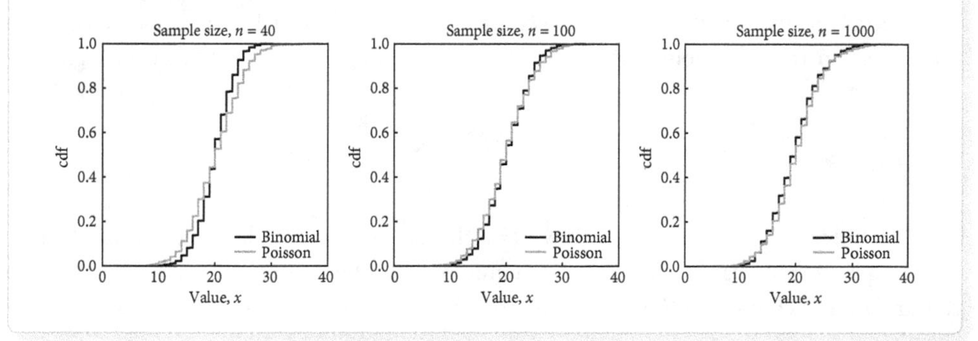

Technical note: Keep in mind that the `binomial` function generates the outcome of `n` trials each with a `p` probability of success. Try to compare outcomes with `sum(np.random.rand(n)<p)`. Are the outcomes different in a substantive way than simply sampling from a binomial? It is worth considering that the binomial distribution is equivalent to running n trials each with a p probability of success and then reporting how many, m, were successful. By definition, $0 \leq m \leq n$. Hence, each trial is successful with probability p. Because `rand` returns uniformly distributed random numbers between 0 and 1,

then `rand < p` is 1 with probability p and 0 with probability $1 - p$. As such, by invoking the `rand` command n times and comparing it to p, it should return a 1 approximately np times; this is, by definition, λ from above. Hence, just as we used the uniform random distribution to generate exponentially distributed numbers, it is also possible to use the same distribution to generate random events that have precisely the same properties as binomial random numbers.

1.5 THE START OF DYNAMICS

The schematics in Figure 1.1 illustrate two distinct mechanisms by which mutant bacteria can increase in number in a population. Via the independent mutation hypothesis, mutations happen rarely during cell division and then selection acts upon them later. Via the dependent mutation hypothesis, mutations only occur when the cell experiences a selection pressure, and in that case a small fraction of heritable cells acquire a mutation. The consequences of these two ideas are examined at length in the main text and then developed as the centerpiece of the homework problems. Yet to get there requires that you develop a dynamic simulation.

Rather than giving away the homework (and the fun involved in doing this yourself), there is a way to start along the path toward dynamics. First, consider the case where mutations are dependent on selection. It should be apparent that manipulating probability distributions as described here can be used to generate a small number of mutants in a population. For example, consider the case where there are $n = 10^5$ cells and $p = 10^{-4}$. In that event, one expects approximately 10 mutational events, which can be generated as follows: `sum(np.random.rand(n)<p)`. Yet the case of the independent selection is more difficult.

As a start to a dynamic model, consider a two-step process. First, a population of cells, with certain features doubles in size. Second, a fraction of the cells changes in some way. Simply to illustrate this point, initialize a 1×5 array with 0.5 in the second entry, e.g.,

```
x=np.zeros(5)
x[1]=0.5
```

Next, double the size of the array. How to do this is up to you. Indeed, doubling an array is perhaps a crude way to simulate a dynamic process, but it provides some intuition to the underlying changes in the system. It also helps illustrate ways to concatenate matrices together, e.g., `np.concatenate([x, x])` Examine the output of `y` and notice that there are now instances of a 0.5 value, in the second and seventh positions. This is a direct result of concatenating the matrices. Now, if the value of 0.5 was some property of a cell, then it is clear that two cells have that same property. If instead one used a value of 1 for a mutant and 0 for a wild type, then it is apparent that the process of cell division (which doubled the number of cells) also doubled the number of mutants. Of course, at this point it would be important to change the property of `y` in the event of a new mutation. In that case, you can use the random number generating methods already described to decide which, if any, of the array elements to change.

Of course, if you want to see what happens in a few instances, consider this loop:

```
x=[1,0]
for i in np.arange(4)
    x=np.concatenate((x,x),axis=None)
```

The result should be a growing list of 0s and 1s:

```
1 0
1 0 1 0
1 0 1 0 1 0 1 0
1 0 1 0 1 0 1 0 1 0 1 0 1 0 1 0
1 0 1 0 1 0 1 0 1 0 1 0 1 0 1 0 1 0 1 0 1 0 1 0 1 0 1 0 1 0 1 0...
```

This kind of approach loses track of the mother-daughter relationships (at least explicitly). But it is possible to modify the arrays and then begin to change both the size and the nature of the population.

How to build models of bacterial growth and mutation is treated in detail in the textbook (and associated homework). From a computational perspective, such models are built around a few simple ideas, including adding elements to an array and changing the value of an array. For example, here are a series of small exercises that illustrate core concepts toward building your own simulation model of bacterial growth and mutation. Type in each and modify them. Soon you may just be ready to tackle the question of whether mutations are dependent on or independent of selection.

CHALLENGE PROBLEM: A Step toward Bacterial Growth

Write a program to generate an in silico population of 100 bacteria, of which $\approx 90\%$ are wild types and the rest are mutants (denote these as 1s and 0s, respectively). Then double the size of this population while retaining the properties of the original population. Finally, switch one element, either 0 to 1 or 1 to 0.

1.6 INFERRING PARAMETERS FROM DATA

Thus far, this laboratory has provided resources for sampling from and manipulating different probability distributions—with an eye toward developing dynamic simulations of growing and mutating bacterial populations. These can be used in a generative sense, as described in the textbook, to compare and contrast the independent and acquired mutational hypotheses. However, there is another question that is relevant to hypothesis testing: how to infer process rates and parameters from data. To tackle such an approach, first download the file `poissdata.csv`, which contains 100 random samples from Poisson distributions with an unknown rate parameter. Or you can enter the following string of numbers into an array. Here it is—exciting, no?

3,4,2,5,2,2,5,0,5,2,4,4,4,1,4,3,3,2,3,2,2,6,3,4,4,5,2,2,5,0,1,2,2,2,4,3,
3,2,4,5,2,4,6,3,5,5,1,3,1,2,2,5,4,8,4,3,5,2,6,3,3,2,3,4,4,3,2,2,3,2,6,2,
2,0,2,5,4,5,4,5,3,9,3,5,2,6,3,5,1,1,2,1,4,2,5,7,4,3,4,4

Although this seems abstract, imagine that these numbers correspond to resistance colonies measured after a Luria-Delbrück (LD) experiment—it turns out that these have features quite distinct from the LD experiments, but they nonetheless provide a good basis for deeper exploration. The remainder of this lab is aimed at estimating the rate parameter, i.e., the unknown λ, from which one could estimate the unknown mutation rate. These steps are the centerpiece of the homework. Hence, it's worthwhile to take some time to understand the *inverse* problem using a simpler example.

The central objective of parameter inference is to try to identify a value (or set of values) that is compatible with observations. The degree of compatibility may depend on one's preference for the unexpected. In practice, most inference approaches try to ask the question: what is the probability that some unknown parameter θ is compatible with the observed data x, or $P(\theta|x)$? Yet, to answer that question, it is often critical to answer a related but different question: what is the likelihood of observing the data x given a parameter θ, or $P(x|\theta)$? These are not the same and, in fact, can be quite different (the literature on false positives in medical testing is an excellent example for study).

In this case, one way to estimate the rate parameter is to use features of the data—and find parameters that are expected to generate similar features. In this case, the pdf for the Poisson distribution is $p(x=k) = \frac{\lambda^k e^{-\lambda}}{k!}$. This means the probability of observing 0 occurrences is $p(0) = e^{-\lambda}$, the probability of observing 1 occurence is $\lambda e^{-\lambda}$, the probability of observing 2 occurences is $\frac{\lambda^2}{2} e^{-\lambda}$, etc. Note that the Poisson distribution is defined over discrete values such that the sum of these must be 1, i.e., $\sum_{k=0}^{\infty} p(x=k|\lambda) = 1$.

One of the features that Luria and Delbrück were interested in was simply the fraction of experiments in which nothing happened—meaning no mutant colonies formed on the agar plates after being exposed to viruses. This feature, the probability of zeros, can be used to infer a rate parameter. In the example, inverting the equation for $p(x=0)$ leads to an estimate of λ based on the data: $\lambda = -\log(P(x=0))$. To calculate the probability of observing 0 from the data, use

```
poissdata = np.genfromtxt('poissdata.csv',delimiter=',')
numberofzeros = np.sum(poissdata==0)
probzero = numberofzeros/len(poissdata)
```

where `sum(poissdata==0)` counts the number of zeros, which then can be used to infer the associated Poisson shape parameter as follows:

- Estimate λ and save the values as `lambda_est`.
- Write code that takes a vector of data as input and outputs the estimated λ based on the number of zero occurrences. Name this function `lambda_estimator_zeros`.

Of note, the data in `'poissdata.csv'` looks like the following, which you should plot to verify:

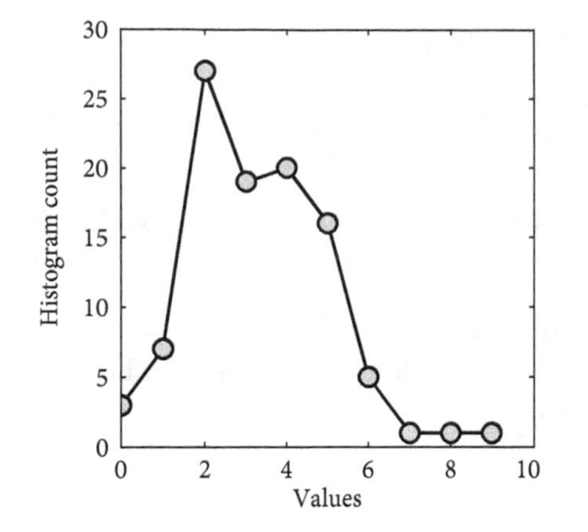

Try to write such a script on your own. However, for your reference, here is one solution script. The danger of course is that, if there are no zeros, then the estimator is undefined. Note that there is another way to estimate λ by using the average value of the output:

```
def lambda_estimator_zeros(x):
    # function lambda_est = lambda_estimator_zeros(x)
    # Estimates the Poisson rate parameter associated with a vector
    # of points x based on the zero
    numberofzeros=np.sum((x==0)*1)
    probzero=numberofzeros/len(x)
    lambda_est = -np.log(probzero)
    return lambda_est
```

What does all of this mean? According to the Poisson distribution, if λ_{true} is the true value, then we should observe an output of 0 a fraction $e^{-\lambda_{true}}$ of time. In the dataset, there are three zeros out of 100 trials. Hence, the observed probability of 0 is $P_{obs}(0) = 0.03$. Hence, our best estimate is $\hat{\lambda} = -log(P_{obs})$, or $\hat{\lambda} = 3.51$. It turns out that is not quite right, yet it is also not surprising; that is, such an output is expected given the true but unknown value λ_{true}. It turns out that the true value is $\lambda_{true} = 3$. But you didn't know that in advance, did you? And that is the point.

Indeed, Luria and Delbrück didn't know what the actual mutation rate was before the experiment (even if they had some idea that it was small, and even some idea of the level of smallness). In the case of any particular estimate, one may ask how confident the estimate of the rate value is. In other words, how often does sampling 100 points from a Poisson distribution with a predetermined rate of λ_{true} lead to similar best estimates of λ_{est}? One way to quantify similarity is to ask whether `lambda_est` lies within the middle 95% of a distribution of estimates obtained from a specified λ_{true}. To get a better sense, let's look at the distribution when we set λ_{true} equal to `lambda_est`. However, in estimating the Poission rate parameter, it is necessary to use a stable statistic—the mean value. Note that the expected value of the Poisson distribution is:

$$\langle x \rangle = \sum_{k=0}^{\infty} kP(k). \tag{1.4}$$

The mean, as it turns out, is simply $\langle x \rangle = \lambda$:

```
numsamps = 10**4
lambdadist = np.zeros(numsamps)
for j in range(numsamps):
    currdata = np.random.poisson(lambda_est,100)
    lambdadist[j] = np.mean(currdata) # Best estimate of $\lambda$
```

Next, plot a normalized histogram of this distribution and address whether `lambda_est` appears to be contained in the middle 95% of the distribution. As seen in the following plot—the answer is yes (as you should have expected):

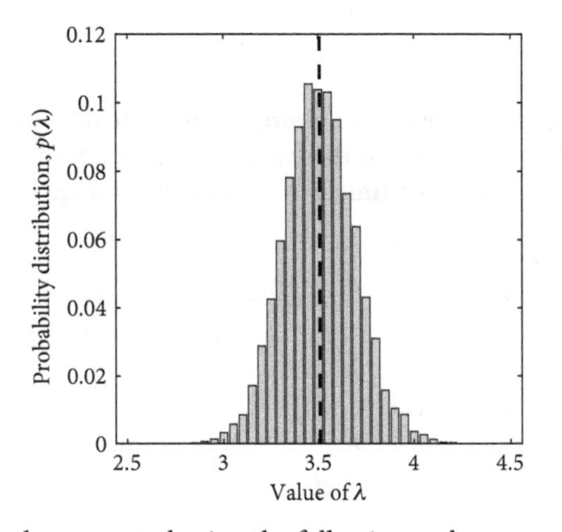

This histogram can be generated using the following code:

```
# Main data goes here
x = np.genfromtxt('poissdata.csv',delimiter=',')
lambda_est = lambda_estimator_zeros(x)

numsamps = 10**4
lambdadist = np.zeros(numsamps)
for i in range(numsamps):
    currdata = np.random.poisson(lambda_est,100)
    lambdadist[i] = np.mean(currdata)

n, bin_edges = np.histogram(lambdadist, bins=30)
bin_probability = n/numsamps
bin_middles = (bin_edges[1:]+bin_edges[:-1])/2.
bin_width = bin_edges[1]-bin_edges[0]
plt.bar(bin_middles, bin_probability, width=bin_width,
        color=[0.75,0.75,0.75],edgecolor='k')
plt.plot([lambda_est, lambda_est],[0, 0.12],'k--',linewidth=3)
plt.ylabel(r'Probability Distribution, $p(\lambda)$', fontsize=14)
plt.xlabel(r'Value of $\lambda$', fontsize=14)
```

Although it is apparent that the first estimate of λ does lie within the center, you can formally identify the bounds to the middle 95% by sorting the distribution as follows:

```
sortedlambdadist = np.sort(lambdadist)
lower025 = sortedlambdadist[int(0.025*numsamps)]
upper975 = sortedlambdadist[int(0.975*numsamps)]
```

Such an outcome might not always be the case. What if you had set $\lambda_{true} = 1$ instead of `lambda_est` (which is equal to 3.51)? To estimate the confidence intervals of the estimated value of λ, one must establish the expected largest and smallest value of λ_{true} with associated distributions that contain `lambda_est` in the middle 95%. We can accomplish this by systematically looping over values of lambda and repeating the analysis above:

```
lambdavec = np.arange(0.5*lambda_est,1.5*lambda_est,0.01)
lower025 = np.zeros(len(lambdavec))
upper975 = np.zeros(len(lambdavec))

for j in range(len(lambdavec)):
    currlambdaset = lambdavec[j]
    poissfitdist = np.zeros(numsamps)
    for k in range(numsamps):
        currdata = np.random.poisson(currlambdaset,100)
        poissfitdist[k] = np.mean(currdata)
    sortedlambdadist = np.sort(poissfitdist)
    lower025[j] = sortedlambdadist[int(0.025*numsamps)]
    upper975[j] = sortedlambdadist[int(0.975*numsamps)]
```

CHALLENGE PROBLEM: Estimating the Confidence in Parameters

In this last problem, plot the lower and upper bounds of the realized values, λ_{obs}, given a range of true values for λ. Then use these forward likelihoods to answer two associated inference problems. First, what is the maximal value of λ with an upper bound below `lambda_est`? Second, what is the minimal value of λ with a lower bound above `lambda_est`? Interpret your findings with respect to the certainty you would have about the underlying value λ_{true} given observations.

SOLUTIONS TO CHALLENGE PROBLEMS

SOLUTION: Properties of Random Distributions

The mean value of the output of `rand` is 0.5 because the values are uniformly spaced between 0 and 1. The variance of a distribution is defined as $\mathrm{Var} = \langle x^2 \rangle - \langle x \rangle^2$, in other words, the expectation of the sampled value squared minus the square of the expected value of the sampled value. For a uniform distribution such that $p(x) = 1$ for $0 \leq x < 1$, the variance is

$$\text{Var}_x = \int_0^1 x^2 p(x) - \left(\int_0^1 xp(x)\right)^2 \tag{1.5}$$

$$= \frac{1}{3}x^3\Big|_0^1 - \left(\frac{1}{2}x^2\Big|_0^1\right)^2 \tag{1.6}$$

$$= 1/3 - (1/2)^2 \tag{1.7}$$

$$= 1/12 \tag{1.8}$$

This result can be verified entering the command `np.var(np.random.rand(10000))`, which returns the variance of 10,000 uniformly distributed random numbers and a number very close to 1/12, or 0.0833. A histogram can be used to visualize the random numbers. As is evident, the shape of the distribution seems largely "flat," though the noise increases with the number of bins. The expected number in a bin is itself a different sampling problem, i.e., a multinomial problem, a topic for a different day. The code to generate the histogram is

```
# Histogram with 25 bins
plt.hist(np.random.rand(10**4),25,\
         facecolor=[0.5,0.5,0.5],edgecolor='k')
plt.xlabel('Value',fontsize=20)
plt.ylabel('Counts',fontsize=20)
plt.title('hist(rand(10**4,1),25)',fontsize=20)
```

This code bins 10^4 random numbers into 25 bins and then visualizes them with labeled axes and a title.

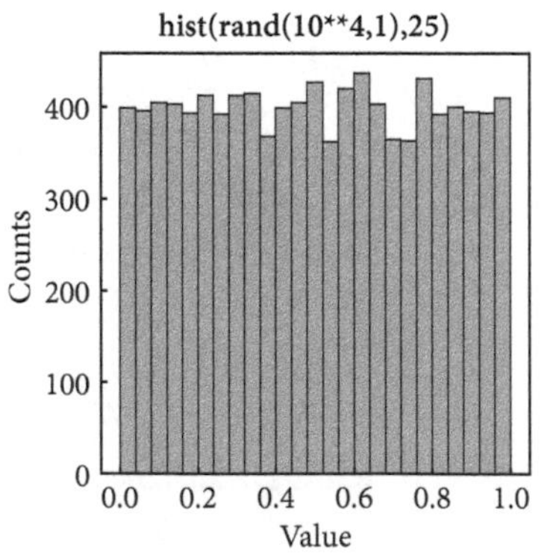

SOLUTION: Random Number Generation

In order to generate this set of random numbers, recall that the `rand` command generates uniformly distributed numbers between 0 and 1. Hence, if you add or subtract a value, then you can shift the range. Moreover, if you multiply the output of `rand` by a constant, you can expand the range. Judicious use of these techniques suggests the following solutions:

- Generate 1000 random numbers equally spaced between 0 and 5: `np.random.rand(1000)*5`
- Generate 1000 random numbers equally spaced between 2 and 7: `np.random.rand(1000)*5 + 2`
- Generate 1000 random numbers equally spaced between −5 and 5: `np.random.rand(1000)*10 - 5`

SOLUTION: Comparing Exponential Random Sampling

The two distributions can be compared using their pdfs (probability distribution functions) or cdfs (cumulative distribution functions). A comparison of the histograms of the pdfs reveals the exponential shape. Transforming the y axis to logarithmic scale would also reveal a linear decline (a bonus challenge for the interested reader). Second, if comparing cdfs, then both distributions map onto each other. In the case of the cdf, the theoretically expected functional form is plotted as an overlay. For plotting pdfs, the following code snippet is useful:

```
N=10**3
my_lambda=1/10
probsamp = np.random.rand(N)
expprobsamp = rand2exp(probsamp,my_lambda)
Pyexprnd = np.random.exponential(1/my_lambda,N)
# Get the histograms
binrange=np.arange(0,105,5)

# Overlay the histograms side by side
plt.hist([expprobsamp, Pyexprnd], binrange,
         weights = [np.ones(N)/N, np.ones(N)/N],
         label=['Custom method','Built-in method'],
         color=['gray', 'black'])
plt.legend(fontsize=12,frameon=False)
```

For plotting cdfs, the following code snippet is useful. In particular, note that this code samples the cumulative distribution at different points to help reveal the overlay of the two approaches to the theoretical distribution.

```
# Generate the random numbers
N=10**3
my_lambda=1/10
probsamp = np.random.rand(N)
```

```
expprobsamp = rand2exp(probsamp,my_lambda)
Pyexprnd = np.random.exponential(1/my_lambda,N)

# Get the empirical cdfs
from statsmodels.distributions.empirical_distribution import ECDF
f1=ECDF(expprobsamp)
f2=ECDF(Pyexprnd)
x1=f1(np.arange(0,max(expprobsamp),max(expprobsamp)/10))
x2=f2(np.arange(1,max(Pyexprnd)+1,(max(Pyexprnd)+1)/10))

# Overlay the cdfs
# First overlay
plt.plot(np.arange(0,max(expprobsamp),max(expprobsamp)/10),x1,'o',
         color=[0.5,0.5,0.5],markeredgecolor='black',markersize=10)
# Different overlay
plt.plot(np.arange(1,max(Pyexprnd)+1,(max(Pyexprnd)+1)/10),x2,'o',
color='white',markeredgecolor='black',markersize=10)
# Theory
x3=np.arange(0,100,0.1)
plt.plot(x3,1-np.exp(-my_lambda*x3),'-', color='black')
plt.legend(['Custom method','Built-in method',\
             r'Theory, $1-e^{-\lambda x}$'], fontsize=12,\
             loc='lower right',frameon=False)
```

SOLUTION: Comparing the Binomial to the Poisson

The comparison is facilitated by generating samples using the following command: `poisssamps = np.random.poisson(n*p,size=numsamps)`. The convergence between binomial and Poisson improves markedly with increasing n, as is apparent in the three sets of cdfs. In each of these cases, 1000 samples were taken given a process that should have an average value of 20. However, when the total number of trials increases from $n = 40$ where $p = 0.5$ to $n = 1000$ where $p = 0.02$, then the convergence to the conditions of the Poisson apply, i.e., large number of trials each with a low probability of success.

SOLUTION: A Step Towards Bacterial Growth

First, generate an array of 0s and 1s, using the `np.random.choice` function, in which $\approx 90\%$ of them are 1s:

```
x = np.random.choice([1,0],size=(100,1),p=[0.9,0.1])
```

Next, try to change a specific type, e.g., by changing it from 0 to 1 or 1 to 0.

```
ind = 20
x[ind]=1-x[ind]
```

Check the value of x before and after to verify that the value did indeed flip. Now double the array in size, copying all elements. There are many ways to do this. In Python, the simplest way is to use the vectorized approach with `np.stack` (though a `for` loop could also work):

```
x.shape
x2 = np.concatenate((x,x),axis=0)
x2.shape
```

This simple command stacks the two column vectors atop each other. In contrast, to copy a column vector so that the resulting array has two (equal) column vectors, use `axis=1`:

```
x.shape
x2 = np.concatenate((x,x),axis=1)
x2.shape
```

Note in the example above that 2 values were given for the size when creating x. This allows Python to differentiate between a column vector and a row vector. If a single number is given, Python will default to creating a 1D array without any notion of rows or columns. In the two examples above, the `shape` attribute provides information on the number of columns and rows. Hence, if x did have information on the wild-type and/or mutant state of many bacteria, then a simple stacking command or direct modification of states could be used to explicitly modify the status of a population at a given time or from one generation to the next. How you use these and other techniques to build a simulation model of cell growth and mutation is up to you!

SOLUTION: Estimating the Confidence in Rates

The plot illustrates the bounds, calculated using a computational approach to generate estimates based on sampled data and then sorting the estimates to identify 95% confidence intervals. Yet to find compatibility requires that we look at what true value of λ could be compatible with the observation of an estimate of 3.51. By looking horizontally across the y axis, one can observe that if the true value had been 2.95, then the value of 3.51 would be a plausible upper limit. Similarly, if the true value had been 3.95, then the value of 3.51 would be a plausible lower limit. This provides a rationale for a confidence interval, as illustrated using the zeros method. See the image below, including a horizontal line at the `lambda_est` value.

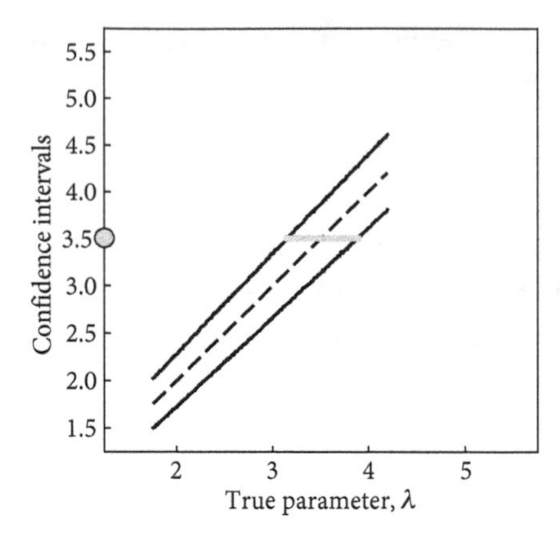

The code to generate this plot is as follows:

```
## Plot commands for estimating lower and upper bounds
plt.plot(lambdavec,lower025,linewidth=3, color='k')
plt.plot(lambdavec,upper975, linewidth=3, color='k')
plt.plot(lambdavec,lambdavec,linewidth=2, linestyle='--', color='k')
tmpi=np.where(lower025 > lambda_est)[0]
llow=lambdavec[tmpi[0]]
tmpi=np.where(upper975 < lambda_est)[0]
lhigh=lambdavec[tmpi[-1]]
plt.plot([llow, lhigh],[lambda_est, lambda_est],
         linewidth=4,color=[0.75,0.75,0.75])
plt.plot(1.25,lambda_est,'ko',\
         markerfacecolor=[0.75,0.75,0.75],markersize=12)
plt.xlim([1.25, 5.75])
plt.ylim([1.25, 5.75])
```

It is possible to use different summary statistics other than the number of zero occurrences. If time permits, repeat this analysis using the mean of the Poisson distribution (equivalent to λ).

Chapter Two

Bistability of Genetic Circuits

2.1 CONTINUOUS MODELS OF CELLULAR DYNAMICS AND GENE REGULATION

The goal of this lab is to simulate the continuous dynamics of living systems. The technical details transcend particular scales, but having a focal example will help structure the effort and provide a basis for adapting lessons learned to different contexts, i.e., across molecules, cells, organisms, populations, and ecosystems. Here the focus is on gene regulation, i.e., how cells modulate the expression of their genes. Modulating gene expression ensures that not all genes are turned on at once. The core components involve the *transcription* of mRNA from DNA and the subsequent *translation* of mRNA into a chain of amino acids, which then fold into mature proteins. These individual events can be represented as processes that modulate the densities of both mRNA and proteins, i.e., the number of molecules per unit volume. For reference, the volume of *Escherichia coli* is approximately 1 μm^3, such that if there is one molecule in a volume of an *E. coli* cell, then that corresponds to a concentration of 1 nM (nanomole). This lab uses units of nM as the basis for analysis.

Simulating the dynamics of mRNA and proteins requires some simplification of the underlying processes. Figure 2.1 shows the core molecular players, described in detail in the main text. Gene regulation denotes the process by which transcription and translation are modified by the state of the system, i.e., by the densities of proteins themselves. This is a form of feedback, where the system state acts to modify process rates. Yet even this level of simplification includes more components and processes than is necessary to build initial models of gene regulation and, in turn, to illustrate core concepts in the simulation of continuous dynamical systems. Instead, the complex cast of players can be reduced to a two-component system involving the densities of mRNA (m) and protein (p). mRNA is transcribed at a rate potentially modified by transcription factors (themselves proteins), then mRNA is translated into proteins, and both mRNA and proteins decay (or are diluted). How does one move from this simplified representation to a representative set of equations and, in turn, a simulation of gene expression dynamics?

To answer this question first requires a mathematical representation of the rules governing changes in concentrations. According to the central dogma of molecular biology, information encoded in genes within DNA is transcribed into mRNA, which is then

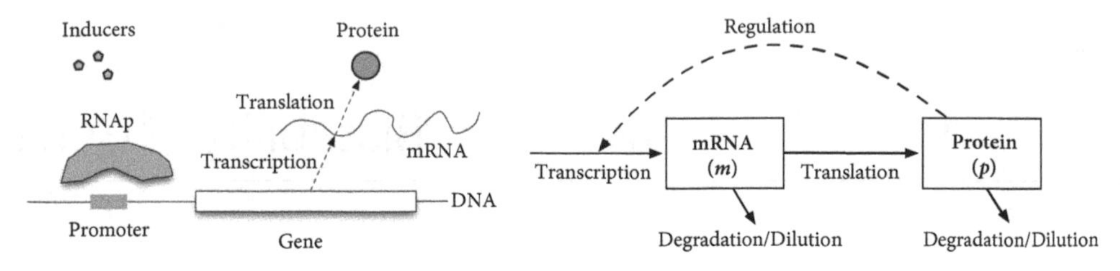

Figure 2.1: Gene regulation in a nutshell. (Left) The molecular players and mechanisms involved in gene regulation. (Right) A schematic view of a model involving the dynamics of mRNA and protein abundances only.

translated into proteins at the ribosome. Here we consider the dynamics of a constitutively active promoter, that is, one that constantly recruits RNA polymerase and produces mRNA of the corresponding gene. By assuming an abundance of RNA polymerase, we can assume the mRNA produce protein copies at a constant per capita rate. This leads to dynamics of mRNA concentration, m, and protein concentration, p:

$$\frac{dm}{dt} = \overbrace{a}^{\text{transcription}} - \overbrace{b_m m}^{\text{degradation/dilution}} \tag{2.1}$$

$$\frac{dp}{dt} = \overbrace{rm}^{\text{translation}} - \overbrace{b_p p}^{\text{degradation/dilution}} \tag{2.2}$$

where a is the rate of mRNA transcription and r is the rate of translation per transcript. The concentration of mRNA and proteins are reduced by degradation and dilution at rate b_m and b_p, respectively. The dilution occurs, in part, because m and p refer to concentrations (number/volume) and the volume of cells grows over time. In total, these represent a coupled system of ordinary differential equations. These equations happen to be linear, but the framework for analyzing them can apply to both linear and nonlinear equations. Compare these equations to Figure 2.1, right. Note that each of the solid arrows corresponds to one term in the coupled system of differential equations. In general, care must be taken to ensure that every process is represented by both additions and losses to the abundances of the relevant systems (e.g., in food webs as modeled in Chapter 11, predation involves both the loss of prey biomass and the gain of predator biomass). The dashed arrow denotes the fact that the rate of transcription might also depend on the levels of protein i.e., $a \rightarrow a(p)$, where $a(p)$ denotes the fact that transcriptional rates depend on the concentrations of proteins; this dependency could be nonlinear.

Given a set of differential equations, how does one solve them? That is to say, given the initial transcript and protein levels in a cell, and rates of transcription, translation, and degradation, what are the expected values of $m(t)$ and $p(t)$, i.e., the time course of mRNA and protein, respectively? Answering this question requires the simulation and analysis of dynamical systems. Core methods for analyzing ordinary differential equations apply whether the underlying model represents gene regulation, neuronal dynamics, or predator-prey dynamics. Simulating this system on a computer requires some way of discretizing continuous time to update the state variables, i.e., in this case $m(t)$ and $p(t)$. This updating is called *numerical integration*; it can be performed using standard functions, e.g., `scipy.integrate.odeint` in Python. The core idea underlying these functions can be understood as follows. First, the instantaneous rate of change is calculated for all variables, i.e., $\mathrm{d}m/\mathrm{d}t$ and $\mathrm{d}p/\mathrm{d}t$. Then, given a suitably small interval of time, the derivative is multiplied by a finite time interval, Δt, to yield an incremental change in the state variables, i.e., Δm and Δp. Given the updated values, the process repeats (Figure 2.2).

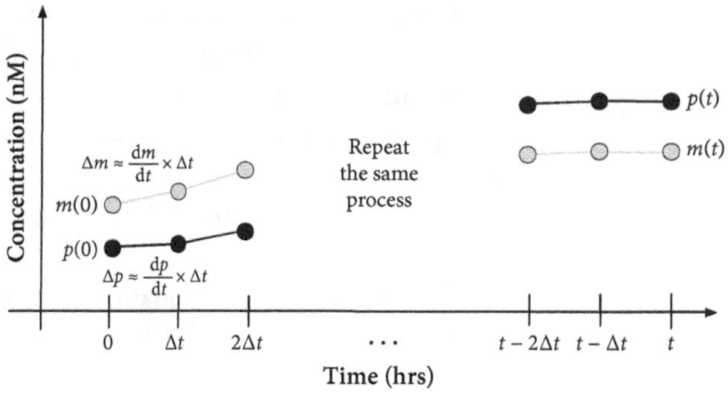

Figure 2.2: Schematic of the Euler integration process.

One major point of caution: Do not actually use a discrete implementation of the Euler integration process; it can be numerically unstable and will often give erroneous results. The function `scipy.integrate.odeint` in Python uses a "higher-order" scheme to ensure that small errors do not compound to ever larger errors in the course of the integration. In practice, we numerically integrate dynamical systems for equations of the form

$$\frac{\mathrm{d}\vec{y}}{\mathrm{d}t} = \vec{f}(\vec{y})$$

where $\vec{f}(\vec{y})$ is a vector of functions. The key idea is that Python numerically integrates a differential equation whose right-hand side must be specified—by you—over a range of time given initial conditions as well as optional parameters. The use of the built-in numerical integration function is described in context below.

But numerical integration is only part of the story. This laboratory introduces a varied toolkit to be deployed in the service of analyzing dynamical systems in the life sciences. The concepts and practical approaches include the identification of fixed points, nullclines, stability analysis, and even a step toward bifurcation analysis.

2.2 SIMULATING COUPLED ORDINARY DIFFERENTIAL EQUATIONS

As noted in Section 2.1, analytically solving equations like those of mRNA and protein dynamics can be difficult or impossible in more complex scenarios. However, numerically solving equations like these with a computer is relatively straightforward. Numerically integrating a coupled system of ODEs is performed via `scipy.integrate.odeint` in Python, which is run with the following command:

```
import numpy as np
import matplotlib.pyplot as plt
from scipy import integrate
t = np.linspace(t0,tf)
x = integrate.odeint(odefun,x0,t,args=(pars,))
```

In this function `t` are points in time, `x` is the solution of the differential equation, `odefun` is the name of the function where you define the set of differential equations, `t0` is the initial time, `tf` is the final time, `x0` is the initial value of `x`, and `pars` is a structure that can be used to pass parameters into the function `odefun`. It is important to note that the variable `x` can represent a vector of variables, e.g., both $m(t)$ and $p(t)$. The last input variable is optional. Hence, without passing any variables, this program can be run as

```
x = integrate.odeint(odefun,x0,t)
```

The differential equation for the protein model is represented with the following function:

```
def proteinproduction(x,t,pars):
    # Returns the instantaneous rate of change given the current time t,
    # state variable x, and parameters (pars)
    m=x[0]
    p=x[1]
    dmdt=pars['a']-pars['bm']*m
    dpdt=pars['r']*m-pars['bp']*p
    dxdt = np.array([dmdt,dpdt])
    return(dxdt)
```

In this function, the set of variables embedded in `x` is converted into variable names that correspond to m and p. This conversion is intentional to help make the code readable, even at a minor cost of some efficiency. Note, that `dxdt` is an array with length equal to the number of variables in the system. In Python, the instantaneous rate of change is a vector; e.g., `dxdt=np.zeros(4)` would be a way to initialize the function in the case of a four-dimensional dynamical system.

But this function is just the start—it describes the instantaneous rate of change, which itself must be used as part of a higher-order Runga-Kutta method to calculate the trajectory (Press et al. 1986). The dynamics can be simulated as follows:

```
import numpy as np
import matplotlib.pyplot as plt
from scipy import integrate
from lab3_functions import proteinproduction

# Parameters
pars={} # Initiate parameter dictionary
pars['a']=10 # nM/hr
pars['bm']=2 # /hr
pars['bp']=1 # /hr
pars['r']=30 # Proteins/(mRNA*hr)

# Initial conditions
x0=[0,0] # Initial values
```

```
t0 = 0 # Time to start
tf = 1 # Final time
t=np.linspace(t0,tf)

# Simulate
x=integrate.odeint(proteinproduction,x0,t,args=(pars,))

# Visualize
fig = plt.figure()
ax = fig.gca()

plt.plot(t,x[:,0],linewidth=3,color='k')
plt.plot(t,x[:,1],linewidth=3,color=[0.5,0.5,0.5])
plt.xlabel('Time (hr)',fontsize=20)
plt.ylabel('Concentration (nM)',fontsize=20)

plt.legend(['mRNA','Protein'],fontsize=20)
ax.tick_params(labelsize=20)
```

The result is the plot on the left, showing the growth of protein and mRNA over 1 hour. Note the units of time (x axis) and concentration (y axis). The plot on the right is the result of the dynamics until the system reaches a steady state (as a self-check, modify the code to simulate out to 24 hours by modifying the final time).

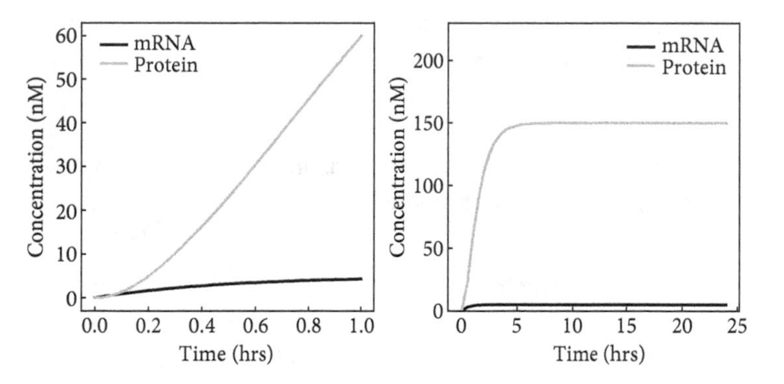

This simulation manages to condense many biological parameters into a single variable—`pars`. There are many ways to organize variables; e.g., in Python, one way is to use "dictionaries." Organizing your variables is a convenient way to group multiple parameters together, both to help read and reuse code and to facilitate passing information from scripts to functions. In Python, a dictionary is a variable that links keys to outputs. The code above introduced four, corresponding to the parameters a, b_m, b_p, and r. It did so by using the `pars[key]` notation, where the variable name is the key and the value is on the right-hand side of the equals sign. In the code block above, these variables have the values $a = 10$, $b_m = 2$, $b_p = 1$, and $r = 30$. However, fields can also denote strings, matrices, and even structures inside structures. Let's give it a try.

CHALLENGE PROBLEM: Organizing Variables

This challenge has two parts: (i) add a new field termed `aRange` to the dictionary `pars` with values from 10 to 50 and a step size of 2; (ii) create a new dictionary called `stats` with three matrices as fields, `tvals`, `mvals`, and `pvals`, each with the number of columns as the length of `aRange` and 100 rows. Initialize all entries to 0.

2.3 QUALITATIVE ANALYSIS OF NONLINEAR DYNAMICAL SYSTEMS

2.3.1 Phase plane visualizations

The previous simulations converged to the values $m(t) \rightarrow 5$ nM and $p(t) \rightarrow 150$. This was expected. In this case, the dynamics converged to a "fixed point", i.e., a combination of state variables in which there is no subsequent change. This fixed point occurs when $\mathrm{d}m/\mathrm{d}t = 0$ and $\mathrm{d}p/\mathrm{d}t = 0$. This is satisfied when $m^* = a/b_m$ and $p^* = \frac{rm^*}{b_p}$, or $m^* = 5$ nM and $p^* = 150$ nM. Yet what happens when the system is not at that fixed point?

To analyze system dynamics requires additional approaches, particularly focusing on transient dynamics. Steps in this direction require that we think differently about how dynamics change in time. Whereas most biologists tend to think about how each variable changes with respect to time, e.g., visualizing many time series stacked one on top of another, an alternative approach is to think about how each variable changes with respect to each other. As such, the dynamics of coupled systems can be represented in terms of their dynamics in phase space, e.g., here the plane represented by $(m(t), p(t))$ where time is implicit rather than explicit. In general, the dimensionality of the phase space is as large as the number of variables. Visualizing such dynamics is possible for low-dimensional systems, like the $m(t)$ and $p(t)$ system studied here.

CHALLENGE PROBLEM: Visualizing Phase Space Dynamics

Write code that visualizes the phase space dynamics of an mRNA-protein system over a 24-hour period with parameters as above, given four different sets of initial

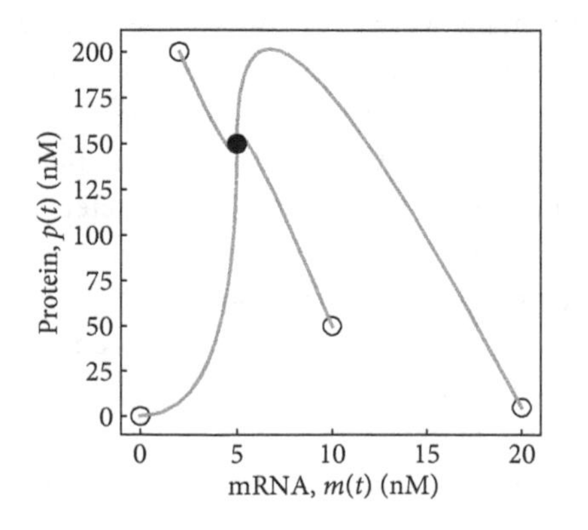

conditions, $(m(0), p(0)) = \{(0,0), (20,5), (2,200), (10,50)\}$. The resulting dynamics should look as shown in the graph, where open circles denote the initial conditions, the gray curves denote trajectories, and the black circle denotes the expected fixed point. (Hint: Use the command `plt.plot` to plot the two output components against one another, corresponding to $m(t)$ and $p(t)$, respectively. In addition, using an array of initial conditions can help with your overlay.)

2.3.2 Nullclines

The next step in qualitative analysis is to understand the trajectory of a system in phase space even when far from the fixed point. To do so, it is helpful to identify the nullclines of the system. A *nullcline* of a particular variables denotes the set of points in phase space where the corresponding dynamical equation is set to 0. Hence, nullclines are associated with each variable, e.g., there are m and p nullclines. For the mRNA-protein model, solving for the nullclines leads to

$$\frac{dm}{dt} = 0 \Longrightarrow m = \frac{a}{b_m} \tag{2.3}$$

$$\frac{dp}{dt} = 0 \Longrightarrow p = \frac{rm}{b_p} \tag{2.4}$$

In other words, there is no change in the mRNA concentration whenever $m(t) = a/b_m$. In addition, there is no change in protein concentration whenever $p(t) = rm(t)/b_p$, i.e., when the ratio of proteins to mRNA is equal to r/b_p. Identifying nullclines also helps to identify the sign of change of variables, e.g., there is usually a sign change associated with the rate of change of one variable when moving from one side of the associated nullcline to the other.

The following commands show how to plot these nullclines in the mRNA-protein space, including the use of anonymous functions:

```
# Define the nullclines
mnullcline = pars['a']/pars['bm']      # A fixed value
pnullcline = lambda p : pars['bp']/pars['r'] * p    # An 'anonymous' function

# Define a vector of protein and mRNA values
pvec = np.arange(0,201,20)
mvec = np.arange(0,9)

# Plot the nullclines
plt.plot(mnullcline*np.ones(np.shape(pvec)), pvec, linewidth=3)
plt.plot(pnullcline(pvec), pvec, linewidth=3)

# Label the plot
plt.legend([r'$\frac{dm}{dt} = 0$',r'$\frac{dp}{dt} = 0$'],fontsize=18)
plt.xlim(min(mvec),max(mvec))
plt.ylim(min(pvec),max(pvec))
```

```
plt.xlabel('mRNA concentration (nMol)',fontsize=20)
plt.ylabel('Protein Concentration (nMol)',fontsize=20)
```

The variable `pnullcline` is a function. The `lambda` keyword means that the input argument p in the assignment is a variable and must be provided as an input when calling `pnullcline`. Note that, like any function, you don't need to call the variable provided to the function by the same name as the local function, and you can even pass in a fixed, scalar value. This code gives the result shown in the graph.

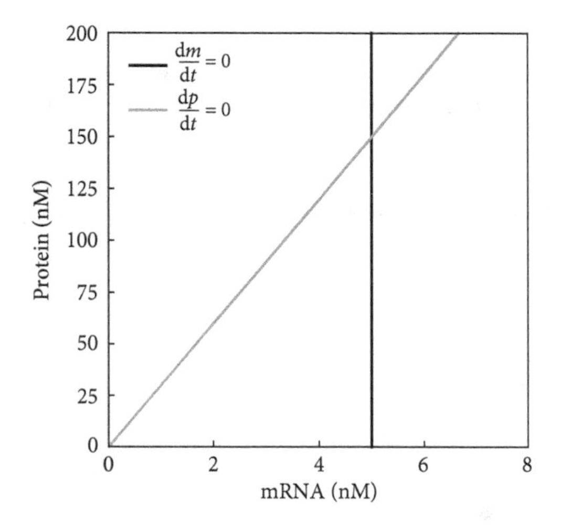

Note that the nullclines cross at precisely one location. At this location both $\mathrm{d}m/\mathrm{d}t = 0$ and $\mathrm{d}p/\mathrm{d}t = 0$. Hence, the intersection of the lines is the equilibrium of the system (m^*, p^*). To visualize the expected local dynamics, it is possible to visualize $\frac{dm}{dt}$ and $\frac{dp}{dt}$ throughout phase space. One way to do so is to plot a mesh of arrows that point in the local direction of change as follows, using the `plt.quiver` function in Python. Note that the following code already presumes the existence of the `mvec` and `pvec` variables as defined above.

```
# Define rate of change
dmdt = lambda pars,m: pars['a']-pars['bm']*m
dpdt = lambda pars,m,p : pars['r']*m-pars['bp']*p
[MM,PP] = np.meshgrid(mvec,pvec)

# Replot the nullclines
plt.plot(mnullcline*np.ones(np.shape(pvec)),pvec)
plt.plot(pnullcline(pvec),pvec)

# Overlay the arrows
plt.quiver(MM,PP,dmdt(pars,MM),dpdt(pars,MM,PP),pivot='mid',
         angles='xy')
```

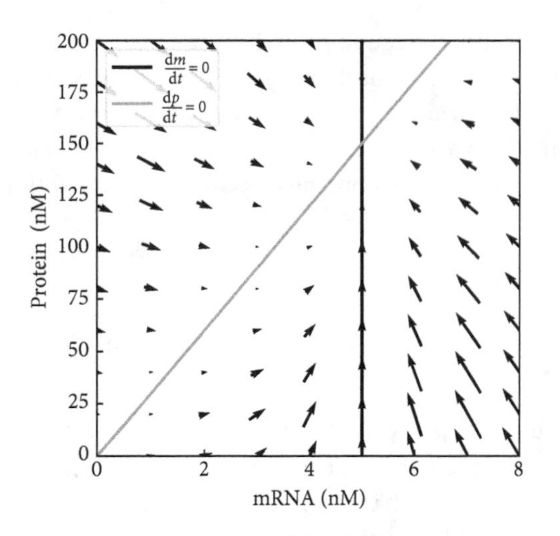

Of course, the nullclines do not necessarily define the actual dynamics. First of all, dynamics can cross nullclines, because they only define where one of the variables (not two variables or all in higher-dimensional systems) has zero rate of change. Instead, the *velocity field* depicts the local direction and magnitude of flow. This velocity field can be used to help interpret anticipated dynamics and even stability, as examined in the next challenge problem.

CHALLENGE PROBLEM: Connecting Dynamics to Nullclines

Simulate Eqs.(2.1)–(2.2) beginning with three initial values in (m, p) space: $(0, 0)$, $(7, 25)$, and $(1, 150)$ over a 0.5-hour period. What do you expect to happen initially? Will both mRNA and protein increase in all cases? Where should the dynamics converge? Then overlay the dynamics onto the velocity field for all three starting conditions, over a 24-hour simulation. To what extent did the nullclines and "quivers" help anticipate trajectories?

2.4 EVALUATING THE LOCAL STABILITY OF EQUILIBRIA

Visually at least, the equilibrium appears to be locally stable, in the sense that trajectories approach it asymptotically. In general, there may be multiple locally stable equilibria in a system. Dynamics approach stable equilibria only if they exist within the associated "basin of attraction," i.e., a region in the phase space—see the textbook for more information. The shape of these basins can be complicated and are not guaranteed to be of a certain size. Hence, to numerically address local stability, it is worth first identifying equilibria either analytically or numerically by solving for the intersection of all nullclines. Then a full accounting of the local stability can usually be assessed through linearization of the dynamical systems near a fixed point. Numerical simulations via small perturbations can be used to verify analytic predictions of local stability or to provide preliminary numerical evidence in the absence of analytical results.

But first we must identify the equilibria of the system. As was suggested above, the fixed points of a system $dx_i/dt = f_i(x, \theta)$ must satisfy $f_i(x^*, \theta) = 0$ for all x_i. Repeatedly solving for variables using the nullclines and plugging into the solutions for the other nullclines solves the algebraic system. This yields an equation for equilibria. But solving such equations is not always guaranteed. By construction, the present case is straightforward:

$$\frac{dm}{dt} = 0 \Longrightarrow m^* = \frac{a}{b_m} \tag{2.5}$$

$$\frac{dp}{dt} = 0 \Longrightarrow rm^* = bp^* \Longrightarrow \frac{ra}{b_m} = b_p p^* \Longrightarrow p^* = \frac{ra}{b_m b_p} \tag{2.6}$$

The present parameters yield $p^* = 150$ nM and $m^* = 5$ nM. Numerical solutions for equilibria can also be used in the event that there are multiple equilibria and solving analytically is difficult, e.g., via the `scipy.optimize.fsolve` command. One approach is to rearrange the solution of the differential equation at equilibrium and provide a guess of the value.

```
from scipy.optimize import fsolve
allnull = lambda p : pars['r']*pars['a']/pars['bm'] - pars['bp']*p
equil_num = fsolve(allnull,100)
```

Here we guess p^* is 100 at equilibrium. By varying our guesses throughout parameter space, it may be possible to identify potential multiple equilibria. More generally, nullclines are useful in identifying how many equilibria exist in the system.

2.4.1 Analytical formalism for linear stability analysis

The local stability of a (nonlinear) dynamical system can be assessed through a series of steps.

- An equilibrium is identified, which we will denote as x^*.
- The nonlinear dynamical system is linearized, in essence by approximating the full nonlinear dynamics of the system via a Taylor expansion near the fixed point. The linearized dynamics are described by the Jacobian (see textbook for full details or refer to Strogatz 1994).
- The stability of the system is identified by finding the eigenvectors and eigenvalues of the Jacobian. The eigenvectors denote those directions in phase space in which a perturbation will grow or shrink exponentially without changing its direction.

In the event that at least one eigenvalue is positive, this implies that small perturbations that have some "projection" in the associated eigenvector direction will grow exponentially. In essence, if one or more eigenvalues has a positive, real component, then the equilibrium is unstable. If all eigenvalues have negative, real components, then the equilibrium is stable. We return to the borderline case of zero eigenvalues later.

How can we turn this relatively small, but potentially opaque, series of steps into practice? That is, how does one calculate eigenvalues and eigenvectors numerically? Recall that the Jacobian is the linearized version of dynamics near the fixed point. The small perturbations can be denoted as $u = m - m^*$ and $v = p - p^*$. Because the original system is linear, this transformation yields a new system of equations:

$$\frac{du}{dt} = a - b_m(u + m^*) \tag{2.7}$$

$$\frac{dv}{dt} = r(u + m^*) - b_p(v + p^*) \tag{2.8}$$

which, replacing the fixed point values using eq. 2.5. and eq. 2.6, yields

$$\frac{du}{dt} = -b_m u \tag{2.9}$$

$$\frac{dv}{dt} = ru - b_p v \tag{2.10}$$

This dynamical system is linear—and would be even if the original had been nonlinear, if one ignores higher-order terms in u and v. As such, we expect the time-dependent solutions to be combinations of exponentials, such that the derivatives with respect to time are just a constant rate factor multiplied by the state variable i.e., $du/dt = \lambda u$ and $dv/dt = \lambda v$, where λ is the unknown eigenvalue.

We can therefore rewrite these dynamics of perturbations as

$$\lambda \begin{bmatrix} u \\ v \end{bmatrix} = J \begin{bmatrix} u \\ v \end{bmatrix} \tag{2.11}$$

where

$$\mathbf{J} = \begin{bmatrix} -b_m & 0 \\ r & -b_p \end{bmatrix}. \tag{2.12}$$

Here $\mathbf{J}$ is the Jacobian matrix. Although the entries in this particular matrix are all fixed parameters, for nonlinear systems these entries also depend on the state variables. In that case, the Jacobian must be evaluated at the fixed point of relevance, i.e., substituting the value of each variable with its corresponding value. The eigenvalues are the solutions to the equation

$$\text{Det}(\mathbf{J} - \lambda\mathbf{I}) \tag{2.13}$$

where λ is an eigenvalue and $\mathbf{I}$ is the 2×2 identity matrix, i.e.,

$$\begin{bmatrix} 1 & 0 \\ 0 & 1 \end{bmatrix}. \tag{2.14}$$

There are as many eigenvalues as there are dimensions in phase space. Hence, for the mRNA-protein systems, there are two eigenvalues. In this case, because of the absence of feedback between proteins and mRNA, the two eigenvalues are $\lambda_1 = -b_m$ and $\lambda_2 = -b_p$. Because both are negative, the system must be locally stable. This result provides the basis for numerical verification in the next section.

2.4.2 Numerical simulations near equilibria

Numerically evaluating the stability of an equilibrium point is possible by simulating the system dynamics near to a fixed point. The following code snippet shows how to do so for the mRNA-protein case.

```
# Find fixed points
pstar=pars['r']*pars['a']/pars['bm']/pars['bp']
mstar=pars['a']/pars['bm']

# Perturb initial conditions
pperturb=pstar*0.01
mperturb = mstar*0.01
x0=[mstar+mperturb,pstar+pperturb]
t0=0
tf=10
t=np.linspace(t0,tf)

# Simulate and visualize just the proteins
x=integrate.odeint(proteinproduction,x0,t,args=(pars,))
plt.plot(t,x[:,1],linewidth=3,color='b')
plt.xlabel('time(hr)',fontsize=20)
plt.ylabel('Protein Concentration (nMol)',fontsize=20)
```

This code snippet perturbed the system 1% above the equilibrium and only plotted the protein dynamics. A similar plot could show that mRNA also returns to the fixed point. By including a second condition just below the fixed point, the code should yield a plot that looks like this:

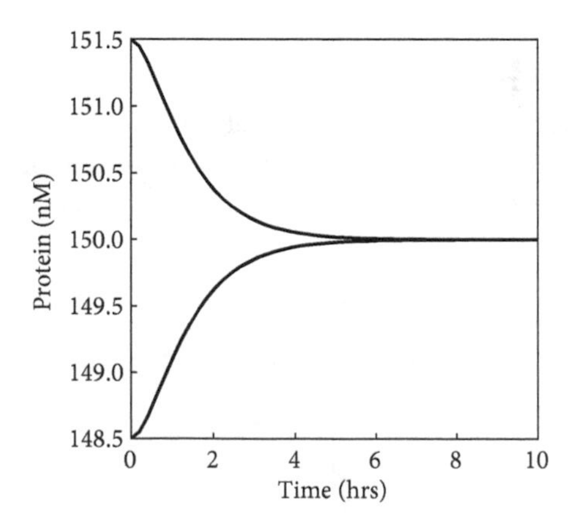

2.5 BISTABILITY AND BIFURCATION DIAGRAMS

This last section is meant to form the final bridge to the problem set in which you will attempt to recapitulate the findings underlying a synthetic toggle switch in *E. coli*. The toggle switch operates given a principle of *bistability*—the idea that a system initialized in different parts of the phase space may end up in different states. These states are also locally stable, i.e., once they end up there, then the system will relax back locally given small perturbations. However, bistability does depend on parameters. Hence, as one changes parameters, the value and even the number of equilibria of the system can change. When a system switches the number and/or stability of its equilibria, we can say the system has undergone a *bifurcation*. These two ideas—bistability and bifurcation—can be assessed numerically. A full analysis will take time, and these concepts recur throughout both this laboratory guide and the main text. Nonetheless, the following sections will help set you on the path to understanding and implementation in practice.

2.5.1 Simulating bistability in one dimension

Gene regulatory dynamics often involve feedback where the levels of proteins modulate the subsequent expression of genes in a network. By way of context, activation (or positive regulation) refers to the case when protein X binds upstream of gene Y and *increases* the production of mRNA and subsequently the protein Y. Similarly, inhibition (or negative regulation) refers to the case when protein X binds upstream of gene Y and *decreases* the production of mRNA and subsequently the protein Y. The toggle switch designed and synthesized by Gardner and colleagues (2000) involves two genes that mutually inhibit one another. This symmetric inhibition can lead to bistability. Hence, even if we were to ignore the mRNA dynamics, it would still require a two-dimensional analysis of the concentration of the two relevant proteins each of which function as transcription factors—a perfect topic for the main text and homework. Yet the concept of bistability can be explored in one dimension. To do so, consider the expression from a gene in which the protein acts as a transcription factor, i.e., modulating the expression of itself.

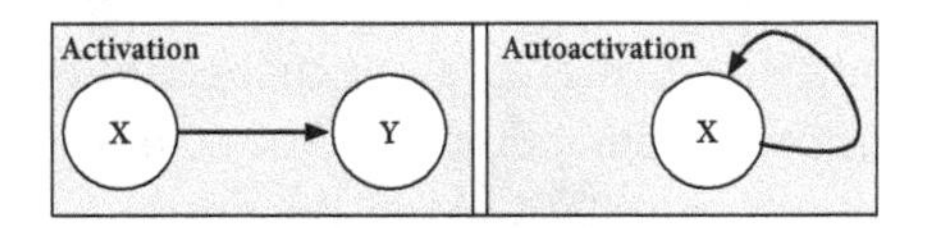

In a self-activating loop, the protein dynamics can be written in the following form:

$$\frac{dx}{dt} = \frac{\beta x^n}{K^n + x^n} - \alpha x \tag{2.15}$$

where x is the density of proteins, β is the maximal production rate, K is the half-saturation constant, α is the decay/dilution rate, and n is a Hill coefficient corresponding to the degree of cooperativity for positive feedback. This model of positive feedback is sufficient, in certain conditions, to generate bistability. The next challenge problem allows you to start the process of exploring bistability on your own. The first step is to construct a function for the dynamical system, as in the following code snippet:

```
def positivefeedback(x,t,pars):
    # Returns the instantaneous rate of change given the current time t,
    # state variable x, and parameters (pars)
```

```
    # for a 1D positive feedback loop
    dxdt = pars['beta']*x**pars['n']/(pars['K']**pars['n'] + x**pars['n'])-\
           pars['alpha']*x
    return(dxdt)
```

CHALLENGE PROBLEM: Simulating a Bistable Feedback Loop

Write your own script to simulate a positive feedback loop and demonstrate that different initial conditions lead to different outcomes. Use the following parameters: $\beta = 100$ nM/hr, $K = 50$ nM, $n = 4$, $\alpha = 1$. If your code works, it should lead to a result like the following:

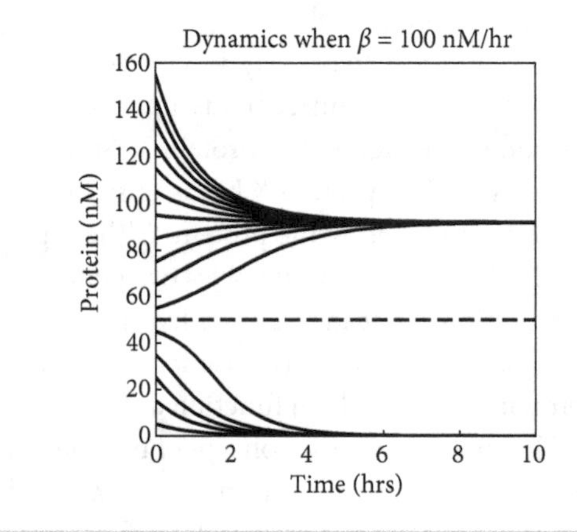

2.5.2 Building visualizations en route to bifurcation diagrams

What happens to the steady state dynamics given changes to parameters in a dynamical system? Will the equilibria change smoothly, or perhaps, quite rapidly? This last section explores this question by systematically varying the degree of positive feedback in the model, i.e., by changing β. Formal bifurcation theory involves more steps. Nonetheless, the flavor of how to generate a bifurcation diagram can be captured in the following figure:

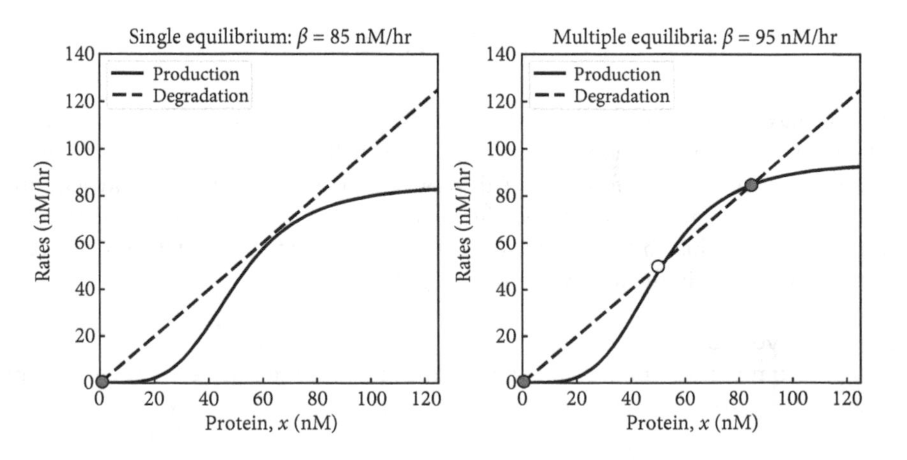

This figure depicts the production and degradation rates as a function of the protein concentration. In both cases $\alpha = 1$ nM/hr, $K = 50$ nM, and $n = 4$. However, the two cases differ in their maximum gene expression rate: $\beta = 85$ nM/hr and $\beta = 95$ nM/hr in the left and right panels, respectively. This relatively small difference (approximately 10% of the total rate) leads to a significant change in output. When $\beta = 85$, then the degradation rate exceeds the production rate for all values of x, i.e., even with positive feedback. Hence, the only potential equilibrium should be when the curves intersect at a null point rather than a nullcline. This equilibrium denotes the Off state when $x^* = 0$, indicated by the black circle. In contrast, when $\beta = 95$, there are three intersections. Each of these represents an equilibrium. Only two of the three are stable (the Off and On states, denoted by black circles) and one is unstable (the intermediate state, denoted by an open circle). This analysis of the dynamical system suggests that small changes in a governing parameter can lead to qualitative changes in the output of the system—a bifurcation! This analysis helps motivate the following framework for thinking about the relationship between dynamical systems and governing parameters. In some sense, one can think of the long-time behavior of a dynamical system as a map from $(\beta, K, n, \alpha) \rightarrow \{x^*\}$ where $\{x^*\}$ denotes a set of fixed points—both stable and unstable. A bifurcation diagram visualizes how this map changes given variation in one (or more) of the input variables. There can be exceptions, e.g., when the steady state does not converge to a fixed point but instead to some form of orbit. Yet, even in that case, the concept of a map holds, albeit to a higher-dimensional space. This may seem abstract, so let's put these ideas into practice.

To start, modify the code from the previous challenge problem so that $\beta = 75$ /hr, $K = 50$ nM, $n = 4$, and $\alpha = 1$ /hr. Using the same code and all other parameters, you should find the following result (the prior result is included for comparison):

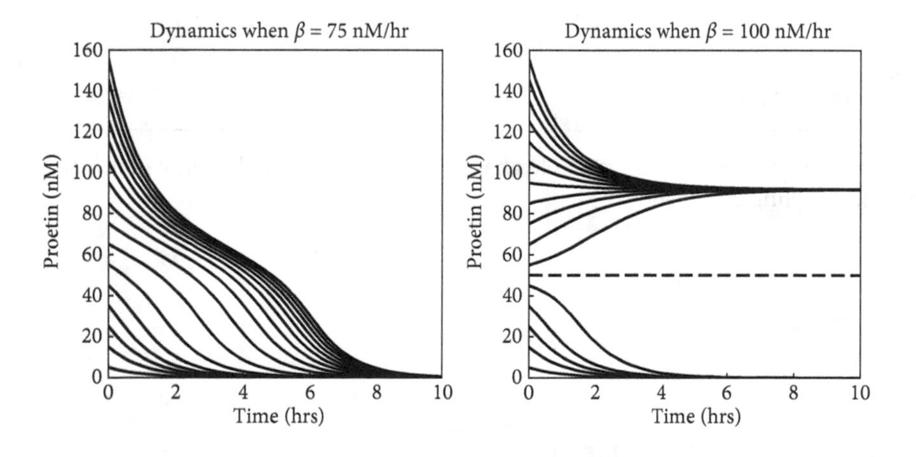

Despite the positive feedback, all trajectories when $\beta = 75$ converge to the Off state, where $x^* = 0$. This difference suggests that the intuition from comparison of the derivatives of the system was correct: a bifurcation in the system dynamics has happened somewhere between $\beta = 75$ and $\beta = 100$.

The homework problems will present a new challenge: to analyze and visualize bifurcations systematically in the context of a synthetic toggle switch—taking these ideas into a systems biology context.

SOLUTIONS TO CHALLENGE PROBLEMS

SOLUTION: Organizing Variables

Challenge (i): Add a new field termed `aRange` with values from 10 to 50 and a step size of 2.

```
# Parameters
pars={} # Initiate parameter dictionary
pars['a']=10 # nM/hr
pars['bm']=2 # /hr
pars['bp']=1 # /hr
pars['r']=30 # Proteins/(mRNA*hr)
# New field value
pars['aRange'] = np.arange(10,50,2)
```

Challenge (ii): Create a new structure called `stats` with three matrices as fields, `tvals`, `mvals`, and `pvals`, each with the number of columns as the length of `aRange` and 100 rows. Initialize all entries to 0.

```
stats = {}
stats['tvals'] = np.zeros( (100,len(pars['aRange'])) )
stats['pvals'] = np.zeros( (100,len(pars['aRange'])) )
stats['mvals'] = np.zeros( (100,len(pars['aRange'])) )
```

Note: You must initialize the dictionary with `dictName = {}`.

SOLUTION: Visualizing Phase Space Dynamics

The solution to this problem involves overlaying multiple phase space dynamics on the same plot. The code snippet below also includes labeling commands.

```
import numpy as np
import matplotlib.pyplot as plt
from scipy import integrate
from lab3_functions import *

# Parameters
pars={} # Initiate parameter dictionary
pars['a']=10 # nM/hr
pars['bm']=2 # /hr
pars['bp']=1 #/hr
pars['r']=30# Proteins/(mRNA*hr)

# Initial conditions
# Array of initial values
```

```
x0_range=[[0,0],
          [20,5],
          [2,200],
          [10,50]]
t0 = 0 # Time to start
tf = 24 # Final time
t=np.linspace(t0,tf)

# Simulate/visualize
for i in range(len(x0_range)):
    x0 = x0_range[i]
    plt.plot(x0[0],x0[1],
             marker='o',
             markersize=12,
             markeredgecolor='k',
             markerfacecolor='w')
    x = integrate.odeint(proteinproduction,x0,t,args=(pars,))
    plt.plot(x[:,0],x[:,1],color=[0.5,0.5,0.5],linewidth=3)

# Visualize the expected fixed point
plt.plot(pars['a']/pars['bm'],pars['r']*pars['a']/(pars['bm']*pars['bp']),
            marker='o',
            markersize=12,
            color='k')

# Labels
plt.xlabel('mRNA, m(t)  (nM)',fontsize=20)
plt.ylabel('Protein, p(t)  (nM)',fontsize=20)
```

SOLUTION: Connecting Dynamics to Nullclines

The dynamics depend on the starting conditions. For both $(0,0)$ and $(7,25)$, the system converges to the equilibrium at $(5,150)$. For $(1,150)$, first proteins decrease and then recover as new mRNA is produced. The realized dynamics are shown below, where the code assumes that you are adding these dynamics on top of the prior visualization commands, using lighter arrows to highlight the vector fields and dashed lines to plot the nullclines.

```
m0_vec = [0, 1, 7]
p0_vec = [0, 150, 25]
t0=0
tf=24
t=np.linspace(t0,tf,200)
for i in range(3):
    x0=[m0_vec[i], p0_vec[i]]
    x = integrate.odeint(proteinproduction,x0,t,args=(pars,))
    plt.plot(x[:,0],x[:,1],linewidth=3,color='r')
```

```
# Don't forget to replot the nullclines and quivers
plt.plot(mnullcline*np.ones(np.shape(pvec)),pvec)
plt.plot(pnullcline(pvec), pvec)
plt.quiver(MM,PP,dmdt(pars,MM),dpdt(pars,MM,PP),pivot='mid',
                    angles='xy')
```

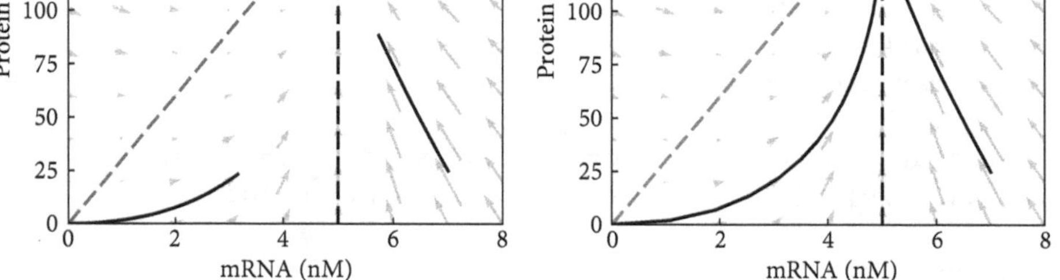

As is apparent, the dynamics have not yet reached equilibrium after half an hour (and appear not to even be that close). The dynamics all converge to the fixed point by 24 hours, and as is apparent, the dynamics can and do cross the nullclines. The other takeaway is that the nullclines provide an indication of local change. Note also that the sign of either the $\dot{m}$ or $\dot{p}$ changes each time the dynamics cross a nullcline.

SOLUTION: Simulating a Bistable Feedback Loop

The following script can be used to simulate a positive feedback loop and demonstrate that different initial conditions lead to different outcomes:

```
# Parameters
pars={} # Initiate parameter dictionary
pars['beta']=100
pars['K']=50
pars['n']=4
pars['alpha']=1

# Set time limits
t0=0
tf=10
t = np.linspace(t0,tf)
# Simulate
x0range = np.arange(5,165,10)
```

```
for i in range(len(x0range)):
    x0 = x0range[i]
    x = integrate.odeint(positivefeedback,x0,t,args=(pars,))
    plt.plot(t,x,color='k',linewidth=3)

# Visualize the split point
plt.plot([t0,tf],[pars['K'],pars['K']],color='k',linestyle='--',linewidth=3)
```

This code does not include the labeling commands. Nonetheless, as is apparent, simulations that begin with low concentrations end up converging to $x^{*,-}=0$, and those that begin with high concentrations end up converging to $x^{*,+}\approx 92$. The meaning of low and high is relative to the point where positive production outpaces degradation; this happens for $x>K$ and $x<x^{*,+}$. The other key point is that dynamics converge and remain at one of the two equilibria. Hence, this positive feedback system is bistable. Note that the high-value equilibrium must satisfy $\frac{100x^4}{50^4+x^4}=x$, which is technically equal to $x^*=91.9643.\ldots$

Chapter Three

Stochastic Gene Expression and Cellular Variability

3.1 SIMULATING STOCHASTIC GENE EXPRESSION

Cellular gene regulation arises from the aggregate outcomes of many individual interactions, e.g., between a transcription factor and a promoter, an inducer and a protein, or an RNA polymerase and a genome sequence. Each of these processes occurs again and again so that the resulting dynamics are often represented as a type of average, or "mean field" behavior. The previous chapter introduced computational methods to simulate deterministic dynamics of gene regulation. Those simulation methods are applicable to other scales, whether of tissues, organisms, populations, or ecosystems. The accuracy of deterministic models improves as the number of processes and number of molecules increases. But one must proceed cautiously, particularly when the numbers are small, which is certainly the case when genes are turned on.

Hence, this laboratory revisits some themes of the previous chapter but with a different goal in mind: simulating stochastic gene expression. Building upon the theory in the textbook, and the previous chapter's development of core programming skills, this lab will help you understand how to (i) translate gene expression processes into dynamic events; (ii) translate dynamic events into a simulation; (iii) simulate stochastic gene expression that recapitulates expected mean field dynamics and that can be compared to measurements. The focal method of this lab is what is commonly known as the Gillespie algorithm (Gillespie 1977). The Gillespie algorithm is a widely used approach developed for simulating chemical reaction kinetics, one event at a time. It can be used in many contexts, including stochastic gene expression. As explained below, the core idea of the Gillespie algorithm is that the state of the cell changes based on individual events. Each event is associated with a process, e.g., degradation or production. These processes have rates, such that it is possible to sample the exact time of the next event from an appropriate exponentially distributed random process. Then, at the next event time, the state of the system changes. For example, if a production event occurs, then the number of proteins goes up by one. In contrast, if a degradation event occurs, then the number of proteins goes down by one. Given an updated system, the process repeats and repeats. A schematic of this process is depicted in Figure 3.1, with multiple realizations of the Gillespie algorithm on the right.

Here then is the target. Consider a cell in which genes are transcribed and then the mRNA translated into protein. We aggregate transcription and translation and assume that

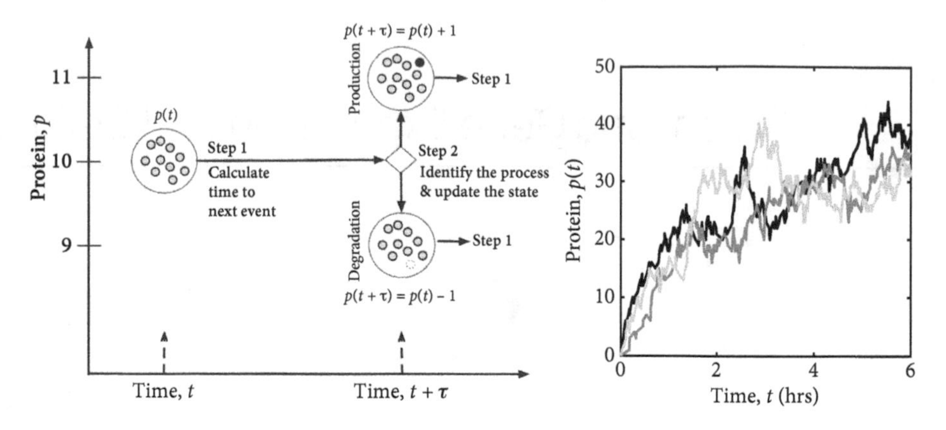

Figure 3.1: Gillespie algorithm, from schematic to application. (Left) A schematic depicting the discrete change in the number of a state variable, in this case proteins. (Right) The result of implementing the Gillespie algorithm for the simplest model of gene expression.

proteins are produced at a rate β, e.g., 50 or 100 nM/hr. The units are important. For example, a single protein in the volume of an *E. coli* cell is approximately 1 nM concentration. Because cells are growing, we assume that proteins decay at an exponential rate of 1 hr^{-1}. This oversimplifies the actual change in concentrations due to cell growth, division, and active degradation but will suffice, for now.

This lab explores the consequences of tracking the dynamics of individual proteins rather than their concentrations. Each section is organized around a series of questions. Will proteins decay to nearly zero, increase without bound, or reach a steady state? How much variation do we expect to find between cells or when examining a cell at different times? In addition, given that stochasticity is now embedded into the core dynamics, we need to revisit the very notion of a "steady state."

The lab focuses on the one-dimensional case, i.e., where the state space is represented by the numbers of proteins. However, the Gillespie algorithm is more general, and can also be used to simulate the stochastic dynamics of multicomponent systems, e.g., those involving mRNA and proteins, or as part of feedback systems with mRNA, proteins, inducers, etc., or eventually as part of entire gene regulatory networks. A secondary goal of this lab is to provide a gateway for simulating seemingly complex models that might not be so complex to simulate after all. The textbook examines the evidence associated with transcriptional bursting, a topic explored in depth in the homework, made possible by the insights and methods provided in the sections to come.

3.2 POISSON PROCESSES: FINDING THE TIME OF THE NEXT EVENT

3.2.1 Waiting time theory

How long does it take for the next production event to take place if the *rate* is b? This rate has units of inverse time, e.g., hrs^{-1}. There tends to be significant misunderstanding about what such units mean. One way to think about them is that the inverse of the rate is a characteristic time, approximately equal to that of the time between random events. So events with a rate of 4 hrs^{-1} occur about every 15 minutes, whereas events with a rate of 10 hrs^{-1} occur about every 6 minutes. Put simply: *Higher rates imply shorter intervals between events.*

Formally, this rate should be thought of as a *probability per unit time* that an event occurs. The probability that an event occurs in some small interval dt is $b \times dt$. The probability that an event does not occur is $1 - b \times dt$. For some, you might recognize such events as arising from a Poisson process. With rates defined, the question we want to ask is, what is the probability that the event takes place at some time t from now? For example, if an event occurs at a rate of 5 per hour, then what is the probability that the *next event* occurs precisely 0.2 hours, 0.4 hours, or 2 hours in the future?

For an event to occur at time t for the first time, it should not occur anytime beforehand. This seems obvious, but it is the key to moving from rates to probabilities. Recall that $p(t)dt$ is the probability that the event occurs between time t and time $t + dt$ where dt is some small time interval, e.g., 0.00001 hours. Therefore, for the event to occur at time t, then it should satisfy the following equation:

$$p(t)dt = \overbrace{(1 - bdt)\dots(1 - bdt)}^{\text{did not occur } \frac{t}{dt} \text{ times}} \times \overbrace{bdt}^{\text{did occur}} \tag{3.1}$$

In words, this means that the event does not occur in any of the $\frac{t}{dt}$ intervals between 0 and t. Then the event does occur with probability bdt in the time interval between t and $t + dt$.

Next, keep in mind that $bdt \ll 1$, i.e., the probability is very small given small time intervals. Hence, we can approximate, $1 - bdt \approx e^{-bdt}$. Combining this together, we find

$$p(t)dt = e^{-bdt\frac{t}{dt}} \cdot bdt$$
$$p(t)dt = e^{-bt} bdt$$

This is the key result. A process that occurs at a rate b leads to *exponentially* distributed events. This is termed a *Poisson* process, i.e., one in which the average time between events is fixed, but the exact time of the next event is unknown and sampled from a memoryless (exponential) distribution.

3.2.2 Some computing

There are multiple ways to generate exponentially distributed random numbers. As explained in the previous chapter, the straightforward approach is to use the built-in random number generator in Python:

```
import numpy as np
import matplotlib.pyplot as plt
b=3
np.random.exponential(1/b)
```

Here the rate is 3 per hour, and the function `np.random.exponential` takes a single input—the inverse of the rate—and returns an exponentially distributed random number. Hence, the input argument to `np.random.exponential` should be the expected mean of the exponentially distributed set of random numbers. Verify that this works by generating 10,000 such random numbers, following the code snippet below to find their mean:

```
tvals = np.random.exponential(1/b,10000)
meant = np.mean(tvals)
```

As you notice, the mean you measure is not *exactly* what you expected given the input rate parameter—1/3? This is expected, given that these are chosen randomly, but the mean is indeed very close to 1/3 and will be ever closer the more samples you take (at least on average). Next up: A challenge problem.

CHALLENGE PROBLEM: Exponentially Distributed Random Numbers

Plot the probability distribution for 10^4 and 10^5 exponentially sampled random numbers. Are the realized distributions in fact exponential? Now change the rate parameter, e.g., setting it to $b = 0.2$, 0.5, 2, or 3. Recall that the `geom_histogram` command is one way to visualize distributions, although there are other options as well. If your code works, the result should look something like the following:

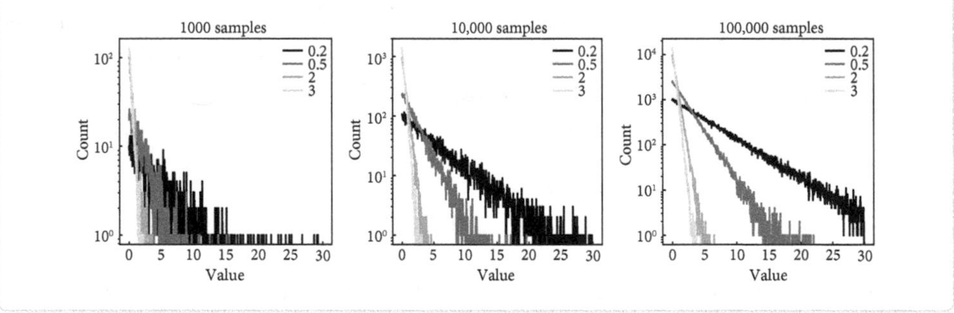

It is also possible to calculate the confidence interval (CI) in the unknown value of b by fitting an exponential to the measured values in x. These CIs are a means to quantify the extent to which an unknown value of $\hat{b}$ could have plausibly generated the observed set of random numbers. It is worth modifying the code snippet to include much smaller sets of random samples—in which case the confidence interval will expand. The code snippet to generate these CIs is as follows:

```
# Number of samples
num = 10**5

bvals = [0.2, 0.5, 2, 3]
b_est = np.zeros(len(bvals))
b_ci = np.zeros((len(bvals),2))
# Loop for each condition of 'b'
for i in range(len(bvals)):
    b = bvals[i]
    x = np.random.exponential(1/b,num)
```

```
    # Best estimates for the mean
    b_est[i] = 1/np.mean(x)
    # Python does not have a built-in way to calculate confidence intervals
    # use bootstrapping
    tmpn=1000
    bootstraps_est = np.zeros(tmpn)
    for j in range(tmpn):
        # Sample from x, allow repeats
        tmpind = np.random.randint(num, size=num)
        bootstrap_x=x[tmpind]
        bootstraps_est[j] = 1/np.mean(bootstrap_x)
    bootstraps_sorted = np.sort(bootstraps_est)
    b_ci[i,0] = bootstraps_sorted[int(tmpn*0.025)]
    b_ci[i,1] = bootstraps_sorted[int(tmpn*0.975)]

# Estimates and CIs contained in b_est
```

By modifying the value of $\hat{b} = 0.2$, 0.5, 2, and 3, the analysis yields the results shown in the following table.

b	$n = 10^4, \hat{b}$	CI	$n = 10^5, \hat{b}$	CI
0.2	0.203	$[0.199, 0.207]$	0.199	$[0.198, 0.200]$
0.5	0.503	$[0.493, 0.513]$	0.501	$[0.500, 0.504]$
2	2.034	$[1.99, 2.07]$	2.00	$[1.99, 2.01]$
3	2.984	$[2.93, 3.04]$	2.99	$[2.97, 3.01]$

3.3 A THEORY OF TIMING GIVEN MULTIPLE STOCHASTIC PROCESSES

The prior section delineates the relationship between the rate of a Poisson process and the timing of the next event. But what happens if more than one kind of event can take place independently, each with its own characteristic rate? What then is the expected time before any event of either kind takes place? This section addresses the issue of identifying the time before *either* process 1 occurs *or* process 2 occurs. For example, if proteins can be produced or degraded, then seemingly some event will happen faster than when only considering production alone or degradation alone. Consistent with the usage thus far, we assume units of hours for time and hrs^{-1} for rates.

To begin, consider two independent Poisson processes with rates $b_1 = 0.5$ and $b_2 = 2.5$. Given those two rates, what is the expected time until the next event? This question can be answered computationally as a guide to the intuition developed in the main text. One way to answer this question is to sample two exponentially distributed random numbers using `np.random.exponential`. Then the waiting time should correspond to the *smaller* of these two waiting times, as in the following code snippet

```
# Set the rates
b1=0.5
b2=2.5
# Generate the anticipated event times
r1=np.random.exponential(1/b1)
r2=np.random.exponential(1/b2)
# Find the smaller time
etime1 = min(r1,r2)
```

The smaller time stored in `etime1` corresponds to the event that occurs first. The function `min` selects the minimum value from the two exponentially distributed random numbers. As an intuitive check, run the code and then ask yourself: was the number closer to 2 (the inverse of `b1`) or closer to 0.4 (the inverse of `b2`)? In reality, a single event is insufficient to help build up one's intuition. Instead, repeat the above procedure 1000 times, recording the times in an array. Here is one way to do it:

```
etime = np.zeros(1000)
for i in range(1000):
    b1=0.5
    b2=2.5
    r1=np.random.exponential(1/b1)
    r2=np.random.exponential(1/b2)
    etime[i] = min(r1,r2)
```

Note that it is possible to write a version of this code with `if` statements—try to figure this out on your own. Irrespective of how you calculate it, consider the following two questions:

- What is the distribution of the variable `etime`, in other words, the expected time to the first event?
- What is the average time to the first event?

Whether plotting the pdf using the `geom_histogram` command or using the cdf, you will notice that the events appear exponentially distributed. They are! The reason is that rates of events add. In other words, the next event occurs as if there were a single Poisson process with rate $b = b_1 + b_2 = 3$ per hour! This means that the average time is closer to 1/3 of an hour (i.e., faster than either process alone!).

Formally, we can compare the distribution to the expected distribution from a Poisson process with $\lambda = 3$. When comparing continuous distributions, it is often useful to compare the cumulative distributions. If the data of times between events is saved as a variable `etime`, then the following code snippet compares the realized cdf versus the theoretically expected cdf. Define $Q(t)$ as the cumulative distribution function, i.e., the probability that the event happens at any time between $0 \leq t' \leq t$; then

$$Q(t) = \int_0^t dt' p(t') = \int_0^t dt' be^{-bt'} \tag{3.2}$$

$$= -e^{-bt'}\Big|_0^t \tag{3.3}$$

$$= 1 - e^{-bt} \tag{3.4}$$

The empirical cdf can be calculated by sorting the n sampled values and finding the fraction of the values whose timing is less than that, for every point $1 \ldots n$. The following code compares the empirical cdf to the theoretical expectation. Note that this first version of the code does not include comments. Try to figure it out yourself, and then look at the documented version following the figure to unpack the meaning of each section.

```
fig = plt.figure(figsize=(5,5))
ax = fig.gca()
b1 = 0.5
b2 = 2.5
n = 1000
x1 = np.random.exponential(1/b1,n)
x2 = np.random.exponential(1/b2,n)
etime = np.min(np.array([x1,x2]),axis=0)
sortdata = np.sort(etime)
cdfvals = np.arange(n)/n
poiss3cdf = lambda t: 1-np.exp(-(b1+b2)*t)
plt.plot(sortdata,poiss3cdf(sortdata),color=[0.5,0.5,0.5],linewidth=6)
plt.plot(sortdata,cdfvals,color='k',linewidth=2)
plt.plot(1/(b1+b2),poiss3cdf(1/(b1+b2)),marker='o',color='k',
         markerfacecolor='k',markersize=12,linestyle='')
plt.legend(['Theory','Empirical','Mean'],fontsize=12,frameon=False,
           loc='upper right')
plt.xlim(0,2.5)
plt.ylim(0,1)
plt.xlabel(r'Time, $t$',fontsize=14)
plt.ylabel('cdf',fontsize=14)
plt.setp(ax.spines.values(),linewidth=2)
ax.tick_params(labelsize=12,direction='in',width=2)
```

The result of this code should lead to the outcome shown in the graph.

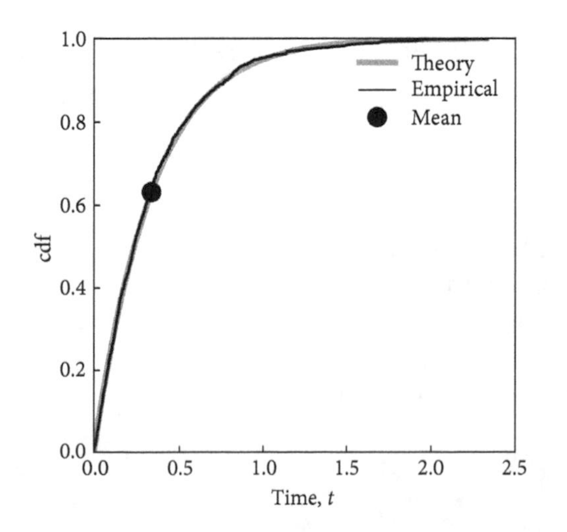

This code has done a few things. Let's diagnose it, including comments for added explanation.

```
# Initialize figure
fig = plt.figure(figsize=(5,5))
ax = fig.gca()

# Generate 1000 random numbers based on the minimum of two processes
b1 = 0.5
b2 = 2.5
n = 1000
x1 = np.random.exponential(1/b1,n)
x2 = np.random.exponential(1/b2,n)
etime = np.min(np.array([x1,x2]),axis=0)

# Sort the event times from smallest to largest
sortdata = np.sort(etime)

# Determine the fraction of the data less than a particular value
# resulting in the cumulative distribution
cdfvals = np.arange(n)/n

# Calculate the cdf of the predicted exponential distribution
poiss3cdf = lambda t: 1-np.exp(-(b1+b2)*t)

# Plot the predicted cdf as a thick gray bar
plt.plot(sortdata,poiss3cdf(sortdata),color=[0.5,0.5,0.5],linewidth=6)
# Plot the empirical distribution
plt.plot(sortdata,cdfvals,color='k',linewidth=2)
# Overlay the location of the expected mean value
plt.plot(1/(b1+b2),poiss3cdf(1/(b1+b2)),marker='o',color='k',
         markerfacecolor='k',markersize=12,linestyle='')

# Add a legend, set axis limits and labels, and increase axis thickness and
# font for readability
plt.legend(['Theory','Empirical','Mean'],fontsize=12,frameon=False,
           loc='upper right')
plt.xlim(0,2.5)
plt.ylim(0,1)
plt.xlabel(r'Time, $t$',fontsize=14)
plt.ylabel('cdf',fontsize=14)
plt.setp(ax.spines.values(),linewidth=2)
ax.tick_params(labelsize=12,direction='in',width=2)
```

With this code complete, it is time for the next challenge.

CHALLENGE PROBLEM: Identity of Random Events

This challenge is centered on solving the following: given two independent Poisson processes operating concurrently, what fraction of times do each of the two processes take place first? That is to say, if $b_1 = 0.5$ and $b_2 = 2.5$, what fraction of events are associated with process 1 and what fraction of events are associated with process 2? Write your own code to try to solve this, leveraging the code already written thus far in the laboratory guide.

This section and challenge problem suggest an alternative method of simulating. Instead of choosing the shortest time, we could first independently sample the time according to a Poisson process with a rate corresponding to any process occurring. In general, the rate of any process occurring is $\lambda = \sum_j \lambda_j$, where λ_j is the rate of a single process, j, and the sum includes all process. Next, we could sample which process occurs where process, j, is chosen with probability $\frac{\lambda_j}{\sum_j \lambda_j}$. This approach is at the core of the Gillespie algorithm.

3.4 GILLESPIE ALGORITHM APPLIED TO A GENE EXPRESSION MODEL

3.4.1 Conceptual framework

The previous section demonstrated how to identify the waiting times between events given two Poisson processes with constant rates. But rates can also depend on the state of the system. For example, in a gene regulatory system, many reactions occur at rates proportional to the concentration of molecular types, e.g.:

- Substrate and enzyme
- Transcription factor and DNA binding site
- Receptor and ligand

There is another difference between the next challenge problem and the prior examples. When an event occurs inside the cell, then the degradation, production, or transformation of a molecule changes the state of the system and therefore has the potential to alter the underlying rates of the governing processes. This feedback is at the core of the Gillespie algorithm.

Returning to the motivating example, consider a cell in which proteins are produced at a constant rate β and decay at a rate α. The abundance of proteins, p, can be represented in terms of a differential equation:

$$\frac{dp}{dt} = \overbrace{\beta}^{\text{production}} - \overbrace{\alpha p}^{\text{decay}}. \quad (3.5)$$

Nonlinear equations of gene regulatory dynamics cannot necessarily be solved analytically, though this particular choice of system does have an accessible closed form solution.

However, it is possible to simulate dynamics arising from such equations via numerical integration. If one numerically integrates this equation given $\beta = 30$ nM/hr^{-1} and $\alpha = 1$ hr^{-1}, the system converges to a fixed amount. Even without integrating, it should be apparent that there is a protein abundance at which production balances decay (a term we use to represent degradation/dilution). That balance happens when $p^* = \beta/\alpha$.

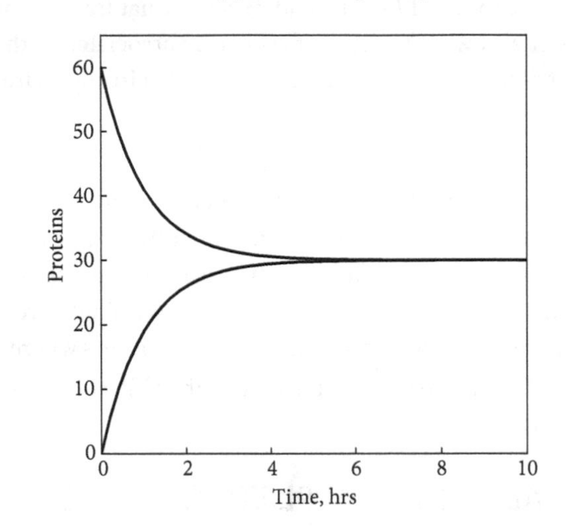

In this graph, the two lines depict the dynamics of protein concentration given a continuous model, starting from two different initial concentrations, 0 and 60 proteins, and an equilibrium of 30.

But are these continuous curves realistic? What happens in a real cell where there can be 0, 1, 2, . . . , 29, 30, 31, . . . of proteins but not 29.231312 proteins? How can a stochastic algorithm capture the random nature of protein production? According to the Gillespie algorithm, the total rate at which any event occurs is the sum of the rates of individual processes:

Production Occurs at a constant rate β, independent of protein concentrations

Decay Occurs at a protein-dependent rate αp

Based on these two processes, first try to sketch the two rates as a function of the protein concentration. Ask yourself: how do they compare and is there a critical protein concentration where the two rates cross (and what exactly might that mean)?

CHALLENGE PROBLEM: State-Dependent Rates

Write code that calculates the probability of decay as a function of the state p. Using this code, calculate the probability of decay at equilibrium, i.e., when $p^* = \frac{\beta}{\alpha} = \frac{30}{1}$. Use code and visualizations to support your argument. If your code is working, it should look like this, depending on whether you visualize from 0 to 60 proteins (left) or from 0 to 300 (right):

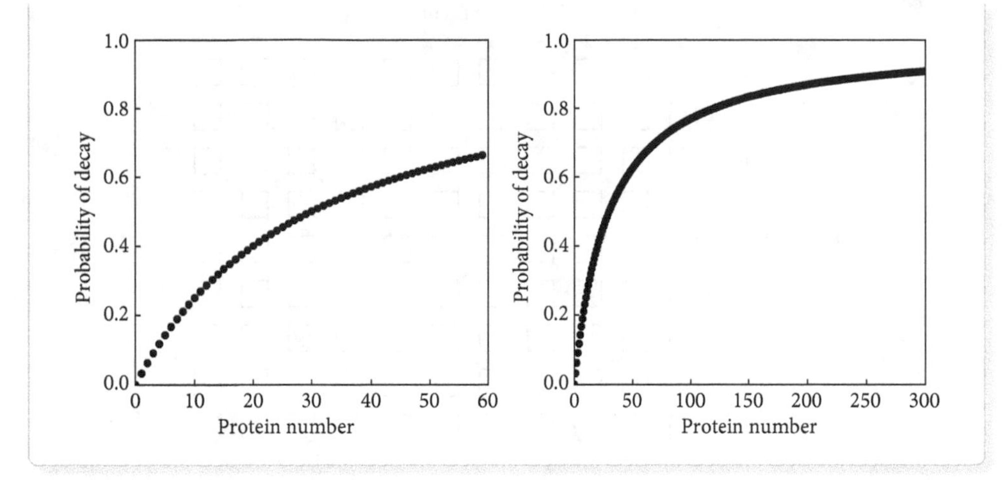

3.4.2 Gillespie algorithm

Before implementing the full algorithm, it is worth discussing the algorithm in a generic form. Computer scientists often call this *pseudocode*. Pseudocode is not written in a specific language but nonetheless lays out a series of steps that your actual code might follow. Here is a pseudocode version of the Gillespie algorithm:

```
specify time range of simulation
set initial time vector and protein numbers vector to 0
while the current time is less than the maximum
        update rate of decay based on number of proteins
        update rate of production
        find the waiting time until the next event
        update the current time based on the waiting time
        decide on which process occurred
        update and store the number of proteins based on selected event
        store the information about the system state at the current time
```

This pseudocode represents an ideal time to pause, strategize, and discuss each step with a classmate or study partner. Try to draw a sketch of what might happen given a cell that initially has four proteins versus one in which there is initially one protein. Is every trajectory the same? Note that in this case the rate of production remains constant irrespective of the current protein level. This need not be the case generally, e.g., as part of gene regulatory networks in which the current expression levels influence new production. The following schematic may be helpful in considering stochastic state trajectories. As you reflect on its meaning, also consider: what would happen if the system started with zero proteins; could the gene still turn on?

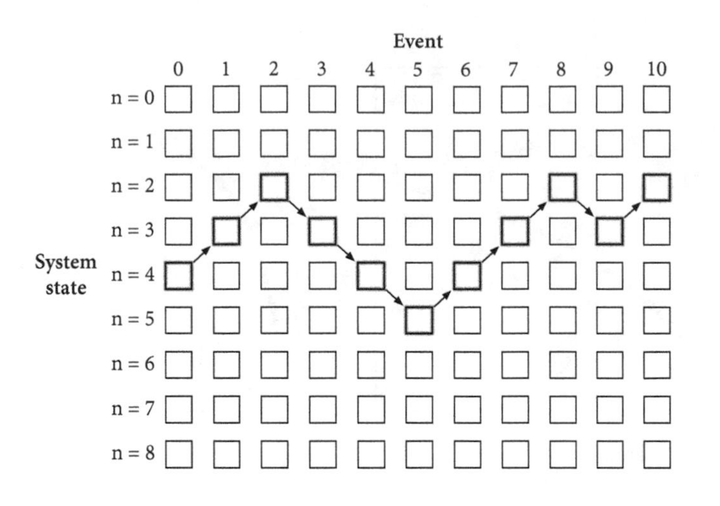

3.4.3 Implementing stochastic gene expression

It is time to build a model of stochastic gene expression using the Gillespie algorithm. The following is one version; feel free to follow it or implement a version on your own! Start a script and enter the following (the comments are optional):

```
# Stochastic gene expression
# Part 1 - initial conditions
p0=0 # Initial proteins
alpha = 1 # Decay rate
beta = 30 # Production rate
currp = p0 # Current proteins
tmax = 6 # Max time
currt = 0 # Current time
tvals = [currt] # Recording of time
pvals = [currp] # Recording of proteins
# End of part 1 - initial conditions
```

With this in hand, now start simulating:

```
# Part 2 - stochastic simulation
while currt<tmax:
    # Calculate rates
    decayrate = alpha*currp
    prodrate = beta
    total_rate = decayrate+prodrate

    # Find waiting time
    deltat = np.random.exponential(1/total_rate) # Find next event
    currt = currt+deltat # Move forward in time
    prob_produce = prodrate/total_rate # Calculate probability of production
```

```
# Update the state
if (np.random.uniform()<prob_produce): # Choose production at random
    currp = currp+1   # Increment number of proteins by 1
else:
    currp = currp-1   # Decrement number of proteins by 1

# Record the state
tvals.append(currt)   # Store the event time
pvals.append(currp)   # Store the protein number
```

That's it! Now plot a stochastic trajectory using the values stored in `pvals`. Three trajectories are plotted below to show some typical variation in the dynamics. The protein number approaches the expected state value of $p^* = \frac{\beta}{\alpha} = \frac{30}{1}$ and then appears to fluctuate about this value. The inherent noise from the sampling process causes the fluctuations around what would be an otherwise stable equilibrium for the analogous ODE dynamics. Here the plot also includes the expected dynamics from the ODE model—it is not a coincidence that the ODE solution runs through what appears to be the center of the stochastic model simulations. As explained in the main text, the ODE solution is the expected mean field (averaged) behavior of the stochastic trajectories.

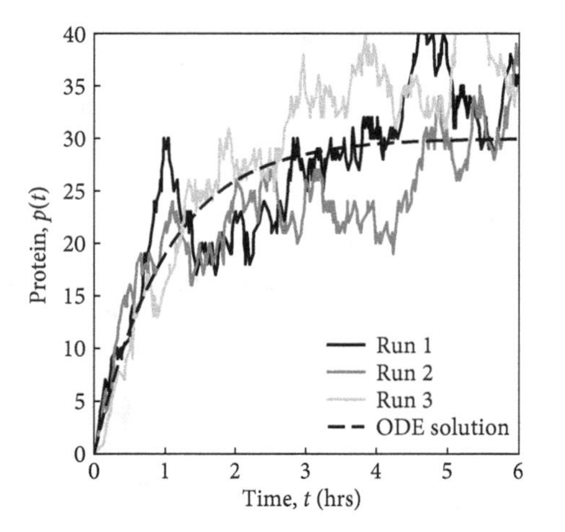

Using this code, try out a few variations:

- Extend or shorten the maximum time.
- Double the production, leaving the decay the same.
- Double both the production and decay rates—does the mean change? If not, do you see any difference?
- Explore the system on your own. . . .

3.4.4 Bonus: Discrete sampling of stochastic trajectories

Comparing stochastic trajectories across time can be difficult because the events occur at random times. It can be useful to sample the trajectory at fixed times. Sampling at fixed times would seem to be intrinsic to the nature of experimental design, but an event-driven

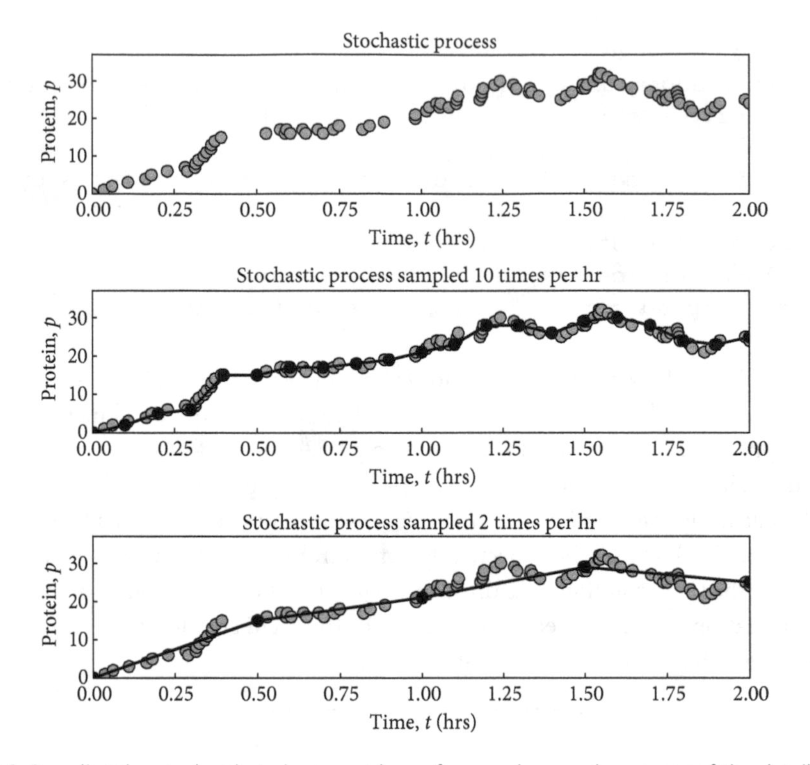

Figure 3.2: Sampling the stochastic trajectory at lower frequencies may lose some of the details of the trajectory.

simulation reports back changes in values at exponentially distributed intervals—these are almost certainly not equally spaced! Hence, the next objective is to learn how to write a function to sample a stochastic trajectory at fixed times. This pseudocode may be helpful:

```
choose vector of times
preallocate vector of times for sampled protein numbers
loop over each time
    find the protein number less than or equal to the current time
    place this value into the vector of sampled protein numbers
end
```

This simple concept is powerful. See, for example, Figure 3.2, in which the results of a single stochastic trajectory are contrasted with the sampled version at a frequency of 10 times and 2 times per hour. Such sampling also makes it possible to compare one trajectory to the next. The figure shows interval sampling of stochastic trajectories, comparing the process (top, circles) with sampled output at 10 times per hour (middle, circles and black line) and 2 times per hour (bottom, circles and line).

CHALLENGE PROBLEM: Discrete Sampling of Stochastic Trajectories

Implement a function to take the output of an event-driven Gillespie algorithm and return the values of the state at discrete or otherwise prespecified intervals. This will be useful in moving from the system perspective to the measurement perspective—see Figure 3.2.

3.4.5 Comparing statistics of the "steady state"

If you have made it this far, you now have a working Gillespie algorithm and a means to sample these stochastic trajectories at fixed intervals. Now you are ready for the next challenge: to compare statistics of the stochastic protein values across time and across distinct trajectories altogether. To do so, first run the Gillespie algorithm initially with 0 proteins for a long time—say, 20 hours. Use the last point as the initial condition for a new simulation. Run the simulation for another 100 hours. Sample this trajectory between 20 and 120 hours for 1000 equally spaced time points (this corresponds to sampling every 6 minutes). What are the mean and variance of the protein numbers across time for this simulation? What is the distribution of protein numbers across time? Now repeatedly simulate the dynamics for 20 hours starting with 0 proteins each time. Do this for 1000 runs. For each simulation store the final protein number. What are the mean and variance of the final protein numbers across the trajectories? Compare this distribution of final protein numbers to the distribution across time that we computed above. The results are remarkably . . . the same! The reasons why are deep, but reflect the *ergodicity* of the problem—the fact that a sufficiently spaced sample of the system's steady state is representative of the entire state space. The following challenge problem formalizes this, though given time constraints you might want to only take on one of the two ways of calculating the steady state. The other remarkable thing is that the distribution itself is a Poisson distribution—for reasons why, see the main text.

CHALLENGE PROBLEM: Steady State Distributions of Stochastic Gene Expression

Develop code that compares and contrasts the output of the long-term sampling of a single trajectory versus sampling of many trajectories at distinct points. Use $\beta = 30$, $\alpha = 1$, and sample at the 20-hour point across 1000 runs or sample every 0.1 hours after reaching the 20-hour point. If your code works, it should look like the graph below.

Notice that the distributions are nearly the same. The main take-away of ergodicity is that, if the dynamics are not expected to change significantly, an experimentalist can choose to obtain statistics by either repeating an experiment or increasing the number of samples of a trajectory. This is not expected to hold when transients are a major factor; hence, we simulated 20 hours after the initial condition before taking data.

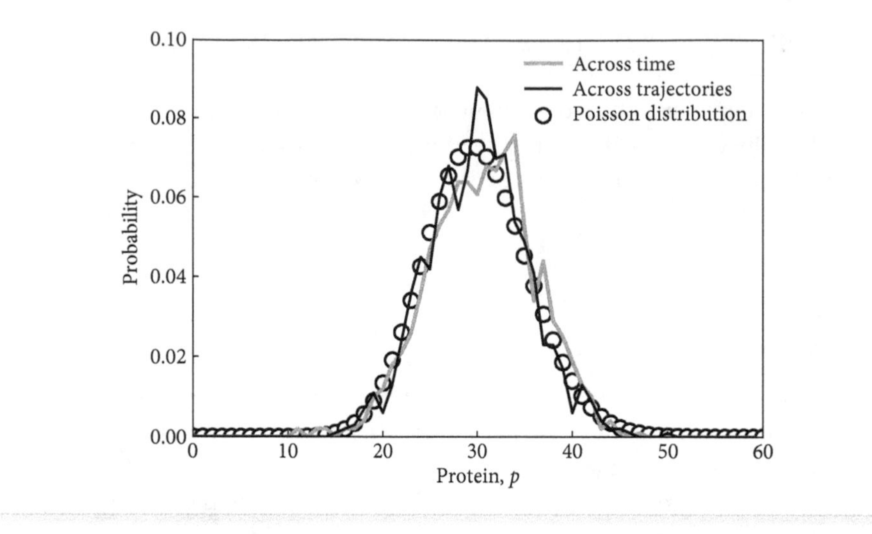

3.5 LOADING AND SAVING DATA

Loading and saving `.csv` files is easy in Python. First, write your trajectory data as a row in the `.csv` file by changing the following strings:

```
import csv
myFile = open('myfilename.csv','w')
with myFile:
    writer = csv.writer(myFile)
    writer.writerow(myarrayname)
```

Then reload it using numpy's built-in reader:

```
x=np.genfromtxt('myfilename.csv',delimiter=',')
```

If the `.csv` file has a header, the `skip_header` argument may denote the number of rows to skip.

SOLUTIONS TO CHALLENGE PROBLEMS

SOLUTION: Exponentially Distributed Random Numbers

The following code generates samples of 10^3, 10^4, and 10^5 random numbers (by varying the variable `num`). One could plot the cumulative distribution function (cdf) or the probability distribution function (pdf). This code uses the pdf, and it is apparent by plotting the histograms on logarithmically scaled y axes that the distributions are in fact exponential. The slopes of the straight lines correspond to the values of b, denoted in the legend. Note that the estimates of the rate parameter correspond closely but not exactly to the generating value. However, the 95% confidence intervals (CIs) are almost certainly going to include the mean—see the textbook for details.

```
import numpy as np
import matplotlib.pyplot as plt

# Setup
num = 1000
bvals = [0.2, 0.5, 2, 3]
b_est = np.zeros(np.shape(bvals))
tmpcol = np.array([[0, 0, 0],
                   [0.25, 0.25, 0.25],
                   [0.50, 0.50, 0.50],
                   [0.75, 0.75, 0.75]])

# Loop for each condition of b
for i in range(len(bvals)):
    # Generate plots
    b=bvals[i]
    # Sample at random
    x=np.random.exponential(1/b,num)
    # Take a histogram with uniform bins
    figTemp = plt.figure(1)
    bins = np.arange(0,30.05,0.05)
    [tmpcount,tmpx,patches]=plt.hist(x,bins)
    figReal = plt.figure(2,figsize=(5,5))
    ax = figReal.gca()
    plt.semilogy(tmpx[:-1],tmpcount,color=tmpcol[i,:],linewidth=3)

# Plot labeling
plt.xlabel('Value',fontsize=20)
plt.ylabel('Count',fontsize=20)
plt.title('{num} samples'.format(num=num),fontsize=20)
plt.legend(list(map(str, bvals)),frameon=False)

plt.setp(ax.spines.values(),linewidth=2)
ax.tick_params(labelsize=12,direction='in',width=2)
```

SOLUTION: Identity of Random Events

The theoretical expectation is that a Poisson event that occurs at a rate b_i among S events occurs before all others a fraction $\frac{b_i}{\sum_i b_i}$ of the time. Hence, in the case where $b_1 = 0.5$ and $b_2 = 2.5$, then the slower process 1 will occur 0.5/3 or 1/6 of the time and the faster process 2 will occur 2.5/3 or 5/6 of the time. The following code snippet shows how to do this, even if you do not know or are not convinced by this argument. The textbook provides a detailed derivation of why this claim holds in general.

```
# Generate random numbers
num=10**3
b1=0.5
b2=2.5
b1_rnd=np.random.exponential(1/b1,num)
b2_rnd=np.random.exponential(1/b2,num)

# Find out which came first
tmpi1=np.argwhere(b1_rnd<b2_rnd)
tmpi2=np.argwhere(b2_rnd<=b1_rnd)
print('Fraction of 1-events that come first is',len(tmpi1)/num)
print('Fraction of 2-events that come first is',len(tmpi2)/num)
```

SOLUTION: State-Dependent Rates

The probability of a decay event taking place before a production event is the ratio of the decay rate to the total rate of all processes. Hence, it is $\alpha p/(\beta + \alpha p)$; that is, a saturating function of p. Note that at equilibrium $\beta = \alpha p^*$, such that by definition, the probability of decay is 1/2 and the probability of production is 1/2, so the system is equally as likely to increase protein levels by 1 as it is to decrease protein levels by 1. The visualization of the probability of decay is enabled by the following code:

```
# Set rates and probability of decay
alpha = 1
r = 30
probdecay = lambda p: alpha*p/(r+alpha*p)
# Plot the probability of decay as a function of proteins
pvec = np.arange(300)
plt.plot(pvec,probdecay(pvec),color='k',marker='.',
          markersize=10,linestyle='')
plt.ylim(0,1)
```

SOLUTION: Discrete Sampling of Stochastic Trajectories

There are many ways to accomplish this goal. Below is a function that implements the pseudocode described above. The function takes the observed time and state values, `t` and `y`, respectively, as input, along with a specific range of times `trange`. The function then returns the times and state values. In essence, the function moves from one sample time point to the next and then scans through the event times until the event time exceeds the sample time. At that point, the value of the state is saved, and then the loop advances. This function retrospectively samples times and states, irrespective of the event timings returned by the Gillespie algorithm. Note that a full code would also deal with the possibility of boundary checking, i.e., ensuring that the sample times and event times are bounded.

```
def sample_traj(t,y,trange):
    # Samples a trajectory t,y at discrete intervals
    t = np.array(t)
    y = np.array(y)
    ts = np.zeros(np.shape(trange))
    ys = np.zeros(np.shape(trange))
    # Initialize
    curt = trange[0]
    ind=np.argwhere(t<=curt)
    curind = ind[-1]
    ts[0]=t[curind]
    ys[0]=y[curind]

    # Scans across the sample interval and then moves the
    # actual dynamics forward until we cross it
    for i in range(1,len(trange)):
        nextt=trange[i]
        while t[curind]<nextt:
            curind=curind+1
        ts[i]=nextt
        ys[i]=y[curind-1]
    return ts, ys
```

With this code in place, a comparison can be made as follows:

```
# Run the Gillespie algorithm
alpha=1
beta=30
tmax=2

# Plot the original
t,y = stochsim_protein([0, tmax],0,beta,alpha)
tmph=plt.plot(t,y,'ko','grey')
```

```
# Sample the trajectory every 0.1 hrs
ts,ys=sample_traj(t,y,np.linsapce(0,tmax,0.1))

# Plot the sampled trajectory
tmph=plt.plot(ts,ys,'ko-',linewidth=3)
```

SOLUTION: Steady State Distributions of Stochastic Gene Expression

The solution to this challenge problem involves running stochastic trajectories, sampling, saving, and repeating. The following code snippet provides sufficient detail to reproduce the main results:

```
# Parameters
beta = 30
alpha = 1

# Long-term trajectory
t, y = stochsim_protein([0,20],0,beta,alpha)
y0 = y[-1]
t, y = stochsim_protein([0,100],y0,beta,alpha)
trange = np.linspace(0,100,1001)
ts, ys = sample_traj(t,y,trange)

# Histogram
py,px = np.histogram(ys,np.arange(0,61))
plt.plot(px[:-1],py/sum(py),linewidth=4,color=[0.5,0.5,0.5])

# Simulate many trajectories
numsamps = 1001
final_y0 = np.zeros(numsamps)
for j in range(numsamps):
    t, y  = stochsim_protein([0,20],0,beta,alpha)
    final_y0[j] = y[-1]

py2,px2 = np.histogram(final_y0,np.arange(0,61))
plt.plot(px2[:-1],py2/sum(py2),linewidth=2,color='k')

# Calculate the distribution in theory
px_theory = np.arange(0,61)
py_theory = stats.poisson.pmf(px_theory,beta/alpha)
plt.scatter(px_theory,py_theory,c='w',edgecolors='k',s=100,linewidth=2)

# Additional plot labels at your discretion
```

Chapter Four

Evolutionary Dynamics: Mutations, Selection, and Diversity

4.1 MODELING EVOLUTIONARY DYNAMICS

The goal of this lab is to learn how to build computational models of evolutionary dynamics. These models date back to the modern origins of the mathematical development of the field. The models—specifically the Wright-Fisher and Moran models—can be adapted to multiple contexts, including changes in the frequencies of a fixed set of genotypes, rapid changes in the genotypes arising via de novo mutation, selection and its consequences, and long-term changes in gene repertoire in a population. To get there requires a number of core techniques, including

- Translating models with stochastic transitions between states
- Simulating genetic drift in a finite-sized population
- Simulating evolutionary dynamics, including the effects of beneficial and/or deleterious mutations

At its core, the lab focuses on computational approaches for simulating Markov processes. A *Markov process* denotes a set of probabilistic rules governing how a system behaves. Markov processes have a special feature: stochastic changes in the state of the system depend on the *current* state and not on prior states. To simulate the dynamics of a cell as a Markov process requires that the frequency (of genes) or abundance (of mRNA, proteins, etc.) are known at a given moment. These abundances then determine the probability of subsequent changes. For example, a Markov process of cellular dynamics might include transcription, protein translation, receptor-ligand binding, and so on. The same principle applies to simulations of populations. For example, specifying the frequency of genotypes at generation $g+1$ in an evolutionary model depends only on the frequency of strains at generation g. In essence, Markov processes are memoryless. In fact, real biological systems can have memory. Nonetheless, it usually possible to expand the number of states within a Markov process to represent whatever memory the system has. The Markov process framework enables the following procedure for building models of living systems—including that of evolutionary dynamics:

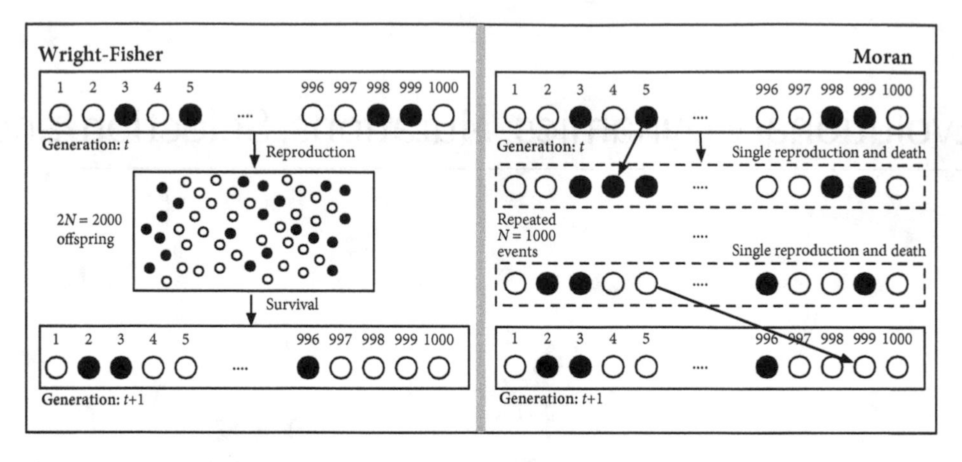

Figure 4.1: Schematic of population genetic models. (Left) The Wright-Fisher model of population genetics as applied to a population of N bacteria. In each generation, each bacterium divides, yielding $2N$ cells. The system is then diluted into fresh media so that only 50% of the cells remain. Hence, after one "generation," the system still has N cells. (Right) Moran model of population genetics as applied to a population of N bacteria. A "generation" includes N birth and death events. In each birth and death event, a random cell dies and is replaced by the offspring of one of the remaining cells.

1. **Initialize state** Set the initial conditions for the model.
2. **Determine the transition probabilities** Based on the current state of the system, determine the probabilities that the system moves to a different state or remains in its current state.
3. **Update the system state** Change the state of the system in a stochastic fashion (i.e., by chance) weighted by the transition probabilities.
4. **Return to step 2 and repeat**

This kind of Markov process can be applied in many contexts. This lab aims to build toward a specific simulation target: evolutionary dynamics of bacterial populations. The primary goal is to develop a Wright-Fisher (WF) model of population genetics to simulate evolutionary dynamics (Wright 1932; Fisher 1958). This model has just a few ingredients—reproduction and survival—as applied to a finite population. Without any other ingredients, these two elements alone can give rise to interesting phenomena. And, with just a bit of work, the WF model can be extended to include selection, mutation, and multiple loci, all of which serve to connect model results directly with experimental evolutionary data. And, with a bit more work, this lab also provides the basis for building a Moran model of evolutionary dynamics (Moran 1958). Both models are shown in Figure 4.1. The differences are subtle and best revealed by coding and doing. Onward!

4.2 TRANSITION MATRICES IN MARKOV PROCESSES

4.2.1 Some preliminary definitions

Markov chains are a special kind of Markov process where discrete transitions between states of a system occur stochastically. The *transition matrix* dictates the probability of a transition between any two states. For example, the graphic below denotes a system with four states (labeled 1, 2, 3, and 4).

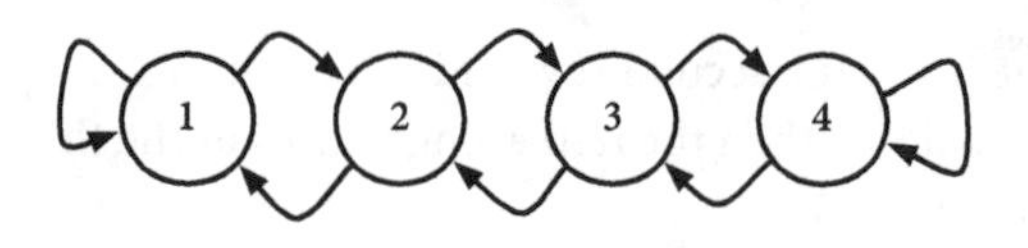

The arrows denote the possible transitions, i.e.,

From state 1 Either to remain in state 1 or to switch to state 2

From state 2 Either to switch to state 1 or to state 3

From state 3 Either to switch to state 2 or to state 4

From state 4 Either to switch to state 3 or to remain in state 4

In order to finalize our description of the system's dynamics, we must also specify the probabilities that the system will make a transition between state i to another state j. As part of this example, assume that all transitions are weighted equally, i.e., they occur with a probability of 0.5 (50%). Hence, we can represent the transition probabilities as a 4×4 matrix, T:

		Current state			
		1	2	3	4
Next state	1	0.5	0.5	0	0
	2	0.5	0	0.5	0
	3	0	0.5	0	0.5
	4	0	0	0.5	0.5

Here the element, T_{ij} is the transition probability from state j to state i; e.g., the first column refers to transitions from the 1 state, the second column refers to transitions from the 2 state, and so on. A few points should be apparent:

- All entries must be nonnegative (i.e., zero or greater).
- The sum of entries in a column must equal 1.

The rationale is that the vertical entries denote transitions from the state in the column to some state in the rows. Because the system must end up in one particular state, then the total probability must equal 1. Later we show that the sum of entries in a row need not equal 1. To start, write your own code to store the transition probabilities in a variable `Tmat`, denoting the transition matrix, so that it returns the following:

```
In [1]: Tmat
Out[1]: array([[0.5, 0.5, 0. , 0. ],
               [0.5, 0. , 0.5, 0. ],
               [0. , 0.5, 0. , 0.5],
               [0. , 0. , 0.5, 0.5]])
```

Note that manipulating a numpy array requires using a "row first" notation. For example,

```
Tmat[2,3]=0.9
```

would assign the value 0.9 to the second row and third column entry of the matrix `Tmat`. Note that, if you were to do so, then the remaining entries in the third column should sum to 0.1 (so that the total sum is 1).

CHALLENGE PROBLEM: Modifying Transition States

Generate a transition matrix by modifying `Tmat` such that the transition probabilities from the second state are 0.2 to the first state, 0.3 to the third state, and 0.4 to the fourth state. Determine the probability that the state does not change and assign this to a new transition matrix—`Talt`.

Critically, this laboratory uses the convention that the columns denote the current state and the rows denote the next state. Some primers use a notation where the rows denote the current state and the columns denote the next state—and then take a transpose when manipulating the transition matrix. Please keep this difference in mind and stick to a convention when writing your code and when thinking about what the transition matrices mean.

4.2.2 Simulating a Markov process

With the transition probabilities in place, it is time to simulate the Markov process. Before writing code, consider a system that starts in state 1. What could happen next given the transition matrix `Tmat`? The key word here is *could*. For example, only one of the following is a viable sequence of states (can you identify it?):

- 1, 1, 2, 3, 4, 3, 4
- 1, 2, 3, 3, 2, 3, 2
- 1, 2, 3, 4, 4, 3, 1

If you are having trouble, recall that only states 1 and 4 have a self-transition, and the system can move, at most, one state away in one single step. Your next challenge: Write a program to simulate dynamics for 50 steps starting at an initial condition $X_{t=0} = 1$. The following code snippet will be helpful, where . . . denotes text to be filled in.

```
Tmat = ...                   # Transition matrix
X0 = 0                       # Initial condition
numgen = 50                  # Number of generations
stochtraj = [X0]             # Initiate the trajectory
currX = X0                   # Current value
for g in range(numgen):
   p_trans = Tmat[:, currX]  # Transition probabilities in the currX column
```

```
nextX = randi_wheel(p_trans)    # Choose a new state based on probabilities
stochtraj.append(nextX)         # Save the state to the trajectory
currX = nextX      # Update the current state
```

All looks fine here, except for the specification of the transition matrix and the selection of the next state. In particular, is there a way to choose the next state without writing a series of bulky `if`, `then`, and `else` statements? Yes, there is. It may help to think of a roulette wheel:

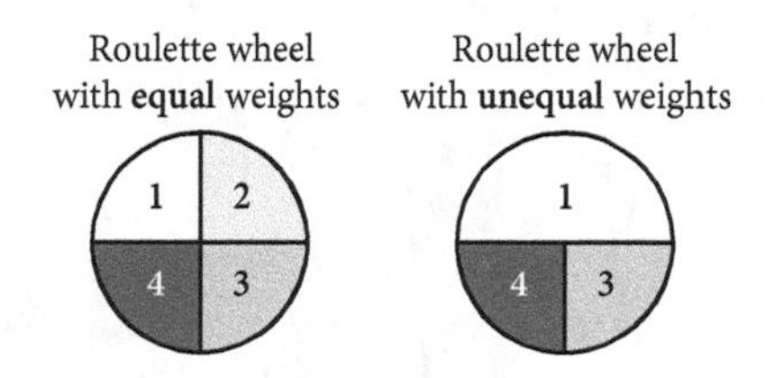

In a (fair) casino (left), the ball is equally likely to land on each number of the roulette wheel. But in an unfair casino (right), or in vintage Bond movies, the ball is unequally likely to land on certain numbers. The transition matrix denotes the *unequal* chances that the system will transition to a given state (right). The transition matrix can include zero values for certain excluded transitions, which means the state is not included in transitions at that given moment. This kind of code block is an example of modular development. As long as the function `randi_wheel` works properly, then this code block represents the entirety of a generalized Markov chain model.

What then does `randi_wheel` do in practice? The function `randi_wheel` takes as input a vector that sums to 1 and returns as output a random index to a state weighted by the values in the vector. In fact, even if the input doesn't sum to 1, then a well-written function should divide by the sum of the weights to ensure that the numbers represent probabilities. In essence, the numbers in the vector represent the relative chance that the state is selected. The algorithm itself draws a random number between 0 and 1 and then checks which state it lands in, i.e., weighted by the wedges of the "biased roulette wheel." Read through and enter in the following function `randi_wheel`.

```
def randi_wheel(weights):
   # Returns a single index returned to the state selected at
   # random specified by the nonnegative values in weights.
   # The function will return a -1 error code if the weights
   # include negative values
     if len(np.argwhere(weights<0)) > 0: # Check for negatives
          x = -1
    else:
          tmp_weights = weights/sum(weights) # Normalize
          tmp_order = np.cumsum(tmp_weights) # Take a cumulative sum
          x = np.argmax(tmp_order >= np.random.rand(1)) # Draw a random number
    return x                         # and find the first ''wedge'' it lands in
```

With this function in place, it is time for a challenge problem.

CHALLENGE PROBLEM: Simulating a Markov Chain

Modify the nearly complete Markov chain simulation code by including the matrix `Tmat` and then plot the dynamics of trajectories. A typical trajectory should look something like this:

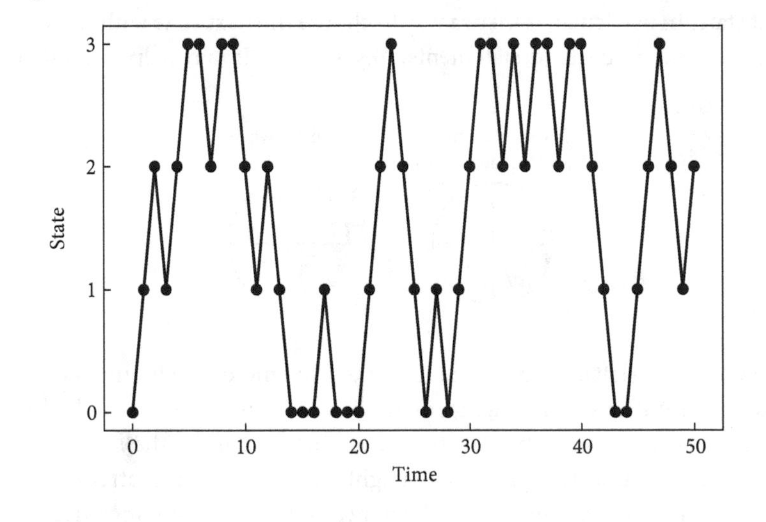

Try this a few times. Is each realization different?

4.2.3 From realizations to the ensemble

The preceding sections enable simulations of a stochastic random process. Before implementing the full WF model, it is essential to understand the difference between an individual simulation and the expected dynamics averaged over many simulations, i.e., within the "ensemble." An ensemble is a collection of many trajectories, each of which share the same transition matrix, T. Indeed, it is not possible to predict the *precise* trajectory of any one realization of the Markov chain. However, it is possible to make accurate predictions of the *average* (or even higher-order properties) of an ensemble of realizations by leveraging the transition probabilities embedded in the transition matrix, T. Before moving ahead, consider changing the simulation of a Markov chain into a function. The benefits of this approach will become apparent in a moment.

```
def markovchain(Tmat,X0,numgen):
   # function stochtraj = markovchain_dyn(Tmat,X0,numgen)
   # Simulates a Markov chain given a specified transition
   # matrix, initial value, and number of generations
   stochtraj = np.zeros(numgen+1)
   stochtraj[0]=X0
   currX = X0
   for g in range(1,numgen+1):
      currfitness = Tmat[:,currX]
```

```
        nextX = randi_wheel(currfitness)
        stochtraj[g] = nextX
        currX = nextX
    return np.array(stochtraj)
```

Now let's return to the transition matrix. To start, denote $\vec{P}_t$ as the probability that the system is in one of the states at time t. The transition matrix holds the key to predicting the probability that the system is in each of the possible states $\vec{P}_{t+1}$ at time $t+1$. For example, in the original example, if the system starts in state 2 at time $t = 0$, then it has an equal chance to transition to state 1 or 3 at time $t = 1$. Hence, the probability of the system state at time $t = 1$ should be $\vec{P}_1 = [0.5\ 0\ 0.5\ 0]$. In general, the transitions in the Markov chain may be written as follows (see the following challenge problem for a visual representation of the equation):

$$\vec{P}_{t+1} = T\vec{P}_t$$

where $\vec{P}_t$ is a vector of probabilities associated with each state at time t. Learning to embrace the transition matrix (or its transpose) takes time, and some hands-on work.

CHALLENGE PROBLEM: Calculating Transition Probabilities

For the matrix T in the image below, calculate the expected probabilities $\vec{P}_{t+1}$ given the two cases of a system certainly in state 2 (case A) and a system that has a 50% chance of being in either state 1 or state 2 (case B).

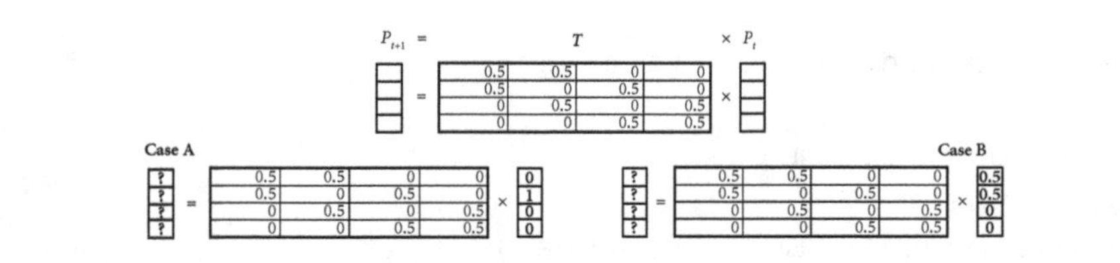

Now repeat the simulations 100 times and store the ensemble of dynamics. Plot the mean value of the state over time. In addition, plot the variance of the value of the state over time. See if you can reproduce something like the following set of images:

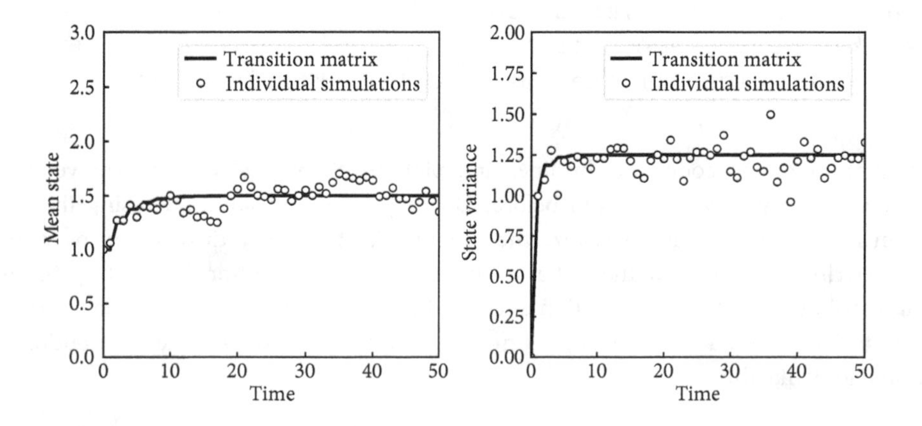

Could you have "predicted" the solid line corresponding to the dynamics simulated via the transition matrix without running thousands (or more) of parallel simulations? Yes, you could have! In fact, the "theoretical" prediction matches the results from averages of individual simulations. The theoretical probability distribution, mean, and variance over time can be determined from repeatedly applying the transition matrix.

```
nvec = np.arange(4)
X0vec = np.zeros(4)
X0vec[1]=1
theormean = np.zeros(numgen+1)
theorvar = np.zeros(numgen+1)
theorp = np.zeros((numgen+1,4))
Xcur = np.array(X0vec)
theormean[0]=np.dot(X0vec,nvec)
theorp[0,:]=X0vec

for i in range(1,numgen+1):
    Xcur = np.matmul(Tmat,Xcur)
    probvec = Xcur
    theormean[i]=np.dot(probvec,nvec)
    secondmoment = np.dot(probvec,nvec**2)
    theorvar[i] = secondmoment-theormean[i]**2
    theorp[i] = probvec

# Simulate the Markov chain, stochastically
numruns = 1000
alldyn = np.zeros((numgen+1,numruns))
for r in range(1,numruns):
    alldyn[:,r] = markovchain(Tmat,X0,numgen)
meanr = np.mean(alldyn,1)
varr = np.var(alldyn,1)

# Overlay the trajectories and the mean
plt.plot(np.arange(0,numgen+1),meanr,color='k',linestyle='',\
         marker='o',markerfacecolor='w')
plt.plot(np.arange(0,numgen+1),theormean, color='k',linewidth=3)
```

Work through the code, line by line, and plot the theoretical curves over your calculated mean and variance from before. Do they agree? In addition, using the transition matrix, we can also visualize the evolution of the entire distribution over time. Next, write code to show the initial evolution of the distribution by plotting the initial distribution and the distribution after the first three steps. (Hint: Use the command `plt.bar(np.arange(4), theorp[i,:])` to visualize the distribution.) It should look like this:

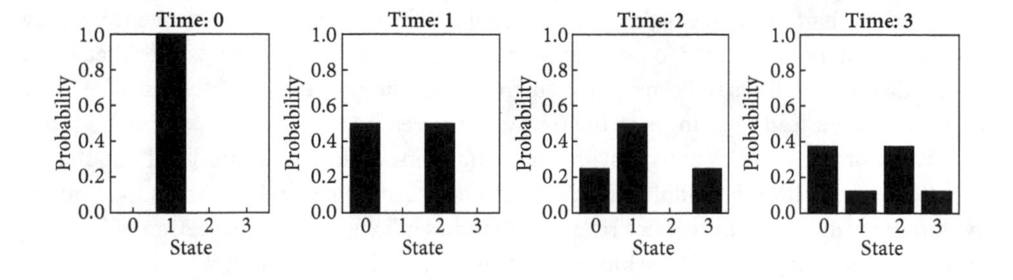

CHALLENGE PROBLEM: Dynamics Toward an Absorbing State

Modify the transition matrix so that

```
In [ ]: Tmat
Out[ ]:
array([[1. , 0.5, 0. , 0. ],
       [0. , 0. , 0.5, 0. ],
       [0. , 0.5, 0. , 0.5],
       [0. , 0. , 0.5, 0.5]])
```

and then simulate the dynamics beginning in state 4. Before implementing this change, ask yourself: what do you expect to happen over the long term? To visualize the dynamics, write a function that plots the distribution $\vec{P}$ for a given t. You can observe the evolution of the distribution by looping through time, plotting the distribution, and calling the `par(ask = TRUE)` command. For example, if your data is stored as `theorp`, then you can watch convergence to an equilibrium distribution over time. Creating quick animations is covered in more detail in Chapters 7–9 of this lab guide. Should the probability be equal for all states, or something else? Run the stochastic model 100 times and store the ensemble of dynamics. Plot the mean value of the state over time and compare with the theoretical expectations.

4.3 THE WRIGHT-FISHER MODEL

It is time to leverage these methods to build the foundational model of evolutionary dynamics: the Wright-Fisher model. As noted at the outset, the WF model represents the dynamics of individuals in a population of fixed size N given non-overlapping generations. In each generation, individuals can reproduce and die. In addition, although individuals may be distinguished by their genotypes or alleles, all reproduction and survival rates are assumed to be equal. This means the WF model is a *neutral* model of evolution. Critically, even though the model is neutral, evolution still happens! Recall the definition introduced in the textbook:

> **Evolution** Any change in the genetic makeup of individuals in a population over subsequent generations

First, if there had been preexisting variation, e.g., different genotypes, then the frequency of each genotype is unlikely to be exactly the same. Second, it is also possible that some of the daughter cells may be mutants. In that case, the process of birth and death (by removal) would lead to changes in the frequency of genotypes. These steps—birth, death, and mutation—are sufficient to provide the foundations of the classic model of population genetics. And, with a few additions, the WF model can be generalized so as to simulate evolutionary dynamics in regimes relevant to studies of experimental evolution of bacteria and yeast, including cases where individual cells vary in their fitness.

4.3.1 Neutral model of evolution: Single locus, two alleles

A simple Wright-Fisher (WF) model treats an evolving population with a single gene with two alleles, A and B, with non-overlapping generations. In short, the population after each generation is obtained by randomly sampling individuals from the previous generation to reproduce. The initial case assumes that there are no fitness differences between alleles. This is a termed a *neutral model*. That is, any A individual has the same probability to reproduce as any B individual. The WF model also provides critical insights into the effect of stochasticity in driving dynamics of evolving, finite populations. To begin, consider a population of size $N = 100$, in which half are A alleles and the other half are B alleles. Given neutral drift alone, to what extent does the number of A alleles change after $g = 100$ generations? There are a few ways to address this question—including both an individual-based and a transition matrix approach—just as in the prior section on Markov chains.

4.3.2 Individual realizations, version 1

Consider a population of size N and recall the schematic of the Wright-Fisher model in Figure 4.1. Notably, the parent of individuals in generation $t + 1$ are random. As a consequence, one way to simulate WF dynamics is to select the parent of each new individual in generation $t + 1$ at random from those parents still present in generation t. The dynamics of a particular WF simulation can look like the following:

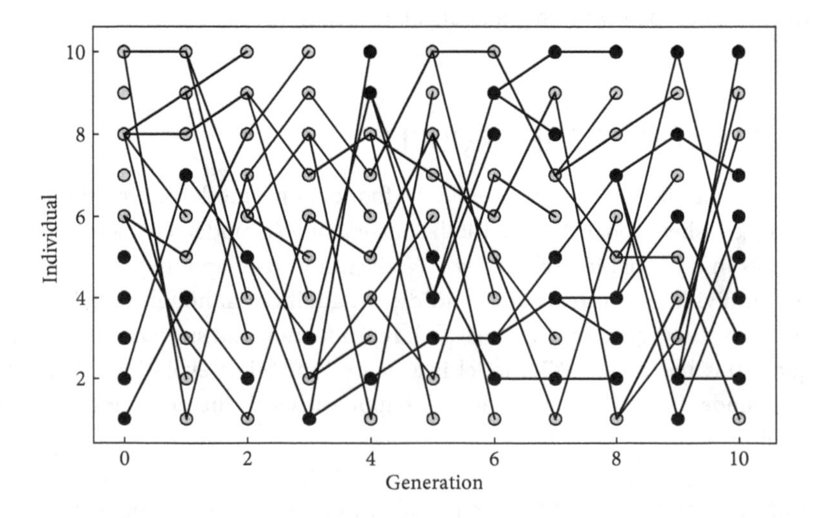

Here lines are used to connect mothers and daughters. Time moves to the right, from generation 0 to 10. But plotting the trace-back networks and thinking about moving backward in time bring up the possibility of analyzing properties of the "coalescent," a bit too advanced for the current lab (Wakeley 2008). This simulation result has many features worth noticing. For example, the visualization spurs the following questions:

- Do all parents have offspring in the next generation?
- Do all surviving offspring in the last generation have different parents from the founding generation?
- What is the chance that a given parent is not chosen to have an offspring in the next generation?

Thinking carefully about each of these questions will help spur intuition for how the WF model works, and how evolution can arise from drift alone.

CHALLENGE PROBLEM: Explicit Wright-Fisher Model

Modify the following code by filling in the sections with . . . to yield an explicit simulation of the WF model.

```
N = 100 # Individuals
numA = 50
numgen = 100
x = np.zeros([numgen + 1, N])
x[0, 0:numA] = 1 # A allele, type 1
x[0, numA:] = 2  # B allele, type 2
fracA = np.zeros([numgen + 1, 1])
fracA[0] = numA/N

for k in  range(3):
    for g in range(numgen):
        for i in range(N):
            tmpid =...      # Find the parent, at random
            x[g+1, i] = ... # Assign the same type to the offspring as the parent
        fracA[g+1] = sum(x[g+1, :] == 1)/N
    plt.hold = True
    plt.plot(range(numgen + 1),fracA, linewidth=2)
    plt.xlabel('Time (generations)',fontsize = 20)
    plt.ylabel('Frequency of A allele',fontsize = 15)

plt.show()
```

If you have this working, then modify the code to visualize multiple trajectories at once. In doing so, notice that some of the trajectories reach the 0 or 1 state—an absorbing state (discussed in Section 4.2), because once the Markov chain reaches there it cannot get out! Once running, the trajectories will look something like this:

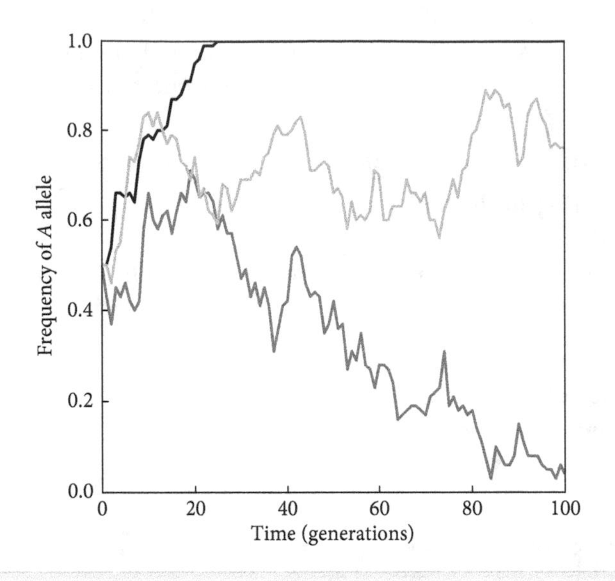

4.3.3 Individual realizations, version 2

This next approach to simulate the WF model ignores the microstates and leverages the symmetry inherent in neutral dynamics. As before, each individual is the same; the difference is only that the fraction of individuals of type A is potentially different from one generation to the next. Hence, the process in version 1 above is equivalent to *sampling from a binomial distribution*. That is, there are N offspring (think of them as trials). For each offspring, there is a $p = n_A/N$ chance that the parent is of type A given that there are n_A individuals of type A in generation t. The probability of selecting k individuals of type A is then

$$p(k) = p^k (1-p)^{N-k} \left(\frac{N!}{k!(N-k)!} \right). \tag{4.1}$$

In other words, a type A individual is selected exactly k times each with probability p, multiplied by the selection of exactly $N-k$ individuals of type B, and finally multiplied by the number of permutations of choosing k individuals out of N (i.e., "N choose k").

CHALLENGE PROBLEM: Binomial WF Simulation

Develop a WF model without keeping track of the individual microstates. Instead, sample directly from a binomial distribution to identify the number of A alleles from one generation to the next. Adapt that code to loop for $g = 100$ generations, storing the state of the population over each generation. Plot the number of A individuals over time. Did the population reach an absorbing state?

The repeatability of neutral dynamics can be assessed by measuring the system's heterozygosity, i.e., the diversity of alleles in the population. Mathematically, heterozygosity is $H = 2p_A p_B = 2p_A(1 - p_A)$, where p_A is the fraction of the population with A alleles and, similarly, p_B is the fraction of the population with B alleles. As a means to explore the code, plot the heterozygosity for a trajectory. You will find that heterozygosity can both increase and decrease. A careful examination will help reveal that heterozygosity is maximized when $p_A = 0.5$, i.e., where there are equal parts of A and B alleles. Further, heterozygosity is minimized when $p_A = 0$ or $p_A = 1$, i.e., the system has reached an absorbing state and therefore there is no diversity in the population. The corresponding heterozygosity dynamics from the previous three trajectories are shown below.

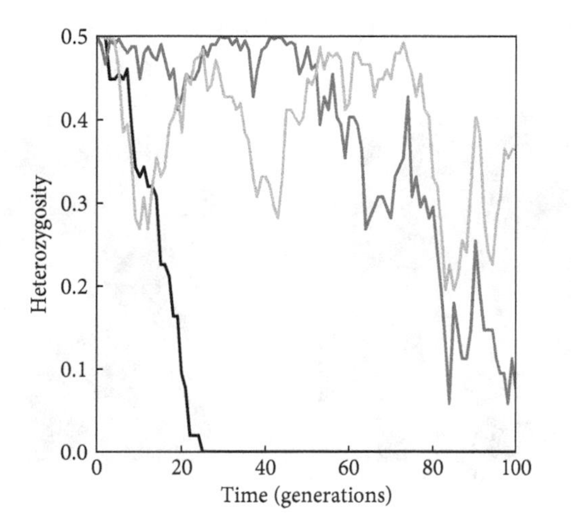

4.3.4 Transition matrix approach

The Wright-Fisher model is an example of a Markov chain. In this case, the states are the number of A alleles in the population. For a system with N individuals, there are $N+1$ states. Hence, when $N = 100$, there are 101 potential states of the system. These correspond to the states in which there are 1, 2, 3, . . . , 99, or 100 A individuals as well as one additional state: corresponding to the case when there are 0 individuals with the A allele. Hence, there are two absorbing states, i.e., where $N_A = 0$ and where $N_A = N$. As in the first example in this laboratory, simulating transitions in a Markov chain requires specifying the transition probabilities. The transition probabilities follow the binomial distribution. That is, for state $i = 1, 2, \ldots, N+1$, the probability that the next generation will have n individuals of type A is given by a binomial distribution with probability $p_i = \frac{i-1}{N}$. As a result, it is possible to construct a transition matrix—exactly analogous to the simple 4×4 state system introduced at the outset—albeit with a biologically motivated (and larger) state space. The following code snippet generates the transition matrix for the WF model using the column notation as before, i.e., each column sum is equal to 1.

```
from scipy import stats
N = 100
Tmat = np.zeros([N+1, N+1])
for jj in range(N + 1):
    currp = jj/N
```

```
    xvec = np.array(range(N + 1))
    Tmat[:, jj] = stats.binom.pmf(xvec, N, currp)
```

To visualize the transition matrix, use the `plt.imshow` command:

```
plt.imshow(Tmat, cmap='jet')
plt.clim([0,1])
plt.colorbar()
```

where the colorbar shows the probability associated with each element of the transition matrix.

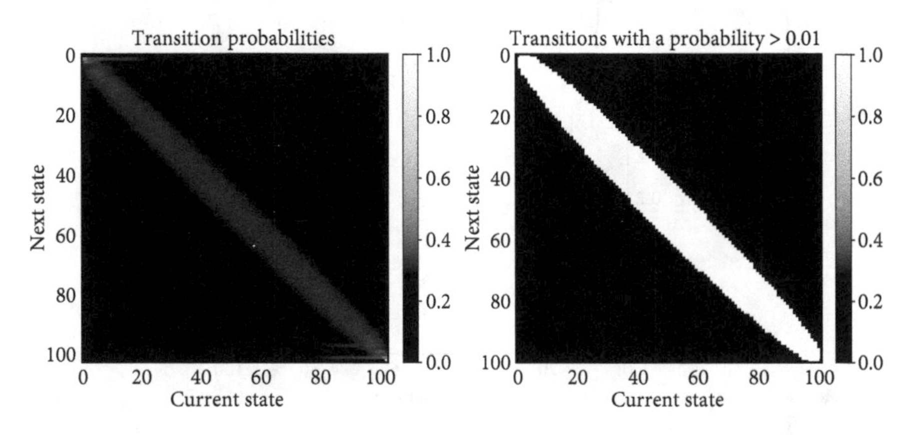

Note that, given the very low probabilities for any particular transition, the right image denotes all those transitions with at least a 0.01 probability—centered around the diagonal. The transition matrix also permits a direct calculation of the theoretical time evolution of the heterozygosity. This involves first solving for the probability distribution at each time given an initial probability distribution.

```
theorp = np.zeros([N + 1, numgen+1])
theorp[int(N/2), 0] = 1
for i in range(numgen):
    theorp[:, i + 1] = np.matmul(Tmat,theorp[:, i])
```

Here the time evolution of the probability distribution follows from repeatedly applying the transition matrix to the vector of initial conditions `X0vec`. The mean heterozygosity follows from

$$\langle H(t)\rangle = \sum_{n=0}^{N} 2\frac{n}{N}\left(1-\frac{n}{N}\right)\mathrm{P}_n(t) \tag{4.2}$$

where $\mathrm{P}_n(t)$ is defined based on the theoretical distribution calculated as `theorp`. The following figure overlays the theoretical and numerical mean heterozygosity dynamics.

It is worth noting that the plot here is on a log-scaled y axis. This is because the heterozygosity is expected to decay exponentially with a time scale of N generations, i.e., $\sim e^{-t/N}$. Note that when $t=N$ the heterozygosity should have declined by a factor of e. Take a closer look at the plot: $0.5e^{-1}$ is approximately 0.18, or precisely where the mean heterozygosity is after 100 generations. A reference code to compare is available below:

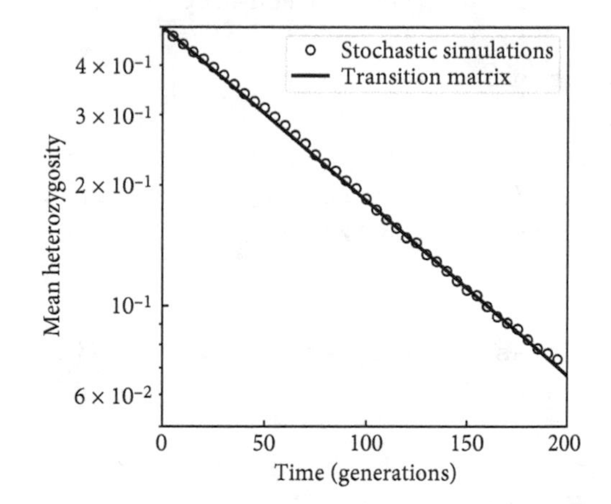


```
# Run the stochastic simulations
numruns = 1000
N = 100
g = 200
allmean = np.zeros((g+1,numruns))
for r in range(numruns):
    allmean[:,r]=WFdyn(N,g)

# Plot the average stochastic models
fig = plt.figure()
ax = fig.gca()
allhetero = 2*allmean*(1-allmean)
meanhetero = np.mean(allhetero,1)
plt.plot(np.arange(0,g,5),meanhetero[0:g:5],
          linestyle = '',
          marker = 'o',
          markerfacecolor = 'w',
          markeredgecolor = 'k')
ax.set_yscale('log')

# Define Tmat = the transition matrix
Tmat = np.zeros([N+1, N+1])
for jj in range(N + 1):
    currp = jj/N
    xvec = np.array(range(N + 1))
    Tmat[:, jj] = stats.binom.pmf(xvec, N, currp)

# Define the prediction p(N) distribution
theorp = np.zeros((N + 1,g+1))
```

```
theorp[int(N/2)+1,0] = 1
for i in range(g):
    theorp[:, i + 1] = Tmat.dot(theorp[:, i])

# Plot the expected H
Avec = np.arange(N+1)/N
heterovec = 2*Avec*(1-Avec)
plt.plot(np.arange(g+1),heterovec.dot(theorp),\
         linewidth=3,color='k')

# Label
plt.legend(['Stochastic simulations','Transition matrix'],\
            fontsize=15)
plt.ylim(0.05,0.5)
```

4.3.5 Clonal interference

If you have reached this point and still want a challenge, consider trying to extend the WF model to include a few more features to simulate "clonal interference," i.e., the simultaneous dynamics of multiple subpopulations. These features could include two loci rather than one, a selective locus, and more alleles (perhaps an infinite number). If you go down this path, you also have to set a mutation rate and consider "linkage." Hence, individuals will be characterized by two states, so there can be far more than two genotypes concurrent at any point. In this way, neutral alleles can hitchhike with selective alleles, and selective alleles can compete with one another. And, if you really want to get fancy, you can make the selective advantages accumulate with time! Soon there won't be much difference between your model and the ones you see in *Nature*, *Science*, and biorxiv!

SOLUTIONS TO CHALLENGE PROBLEMS

SOLUTION: Modifying Transition States

The transition matrix `Talt` can be written as

```
Talt = Tmat
Talt[:,1]=np.array([0.2,0.1,0.3,0.4])
```

or by assigning the transition probabilities directly as follows:

```
Talt = np.array([[0.5, 0.2, 0,   0],
                 [0.5, 0.1, 0.5, 0],
                 [0  , 0.3, 0  , 0.5],
                 [0  , 0.4, 0.5, 0.5]])
```

SOLUTION: Simulating a Markov Chain

The transition matrix should be 4×4 in dimension. Each column denotes an originating state, and the value in the row denotes the probability that a system in the state denoted by the column transitions to the state denoted by the row. Critically, each trajectory is stochastic and therefore is likely to be distinct, particularly once simulated for a sufficiently long time. In addition, your trajectories should never get stuck in one state—such states are termed *absorbing states*, for which there is a way in but no way out.

```
Tmat = np.array([[0.5, 0.5, 0. , 0. ], # Transition matrix
                 [0.5, 0. , 0.5, 0. ],
                 [0. , 0.5, 0. , 0.5],
                 [0. , 0. , 0.5, 0.5]])
X0 = 0                  # Initial condition
numgen = 50             # Number of generations
stochtraj = [X0]        # Initiate the trajectory
currX = X0              # Current value
for g in range(numgen):
    p_trans = Tmat[:, currX]      # Transition probabilities
    nextX = randi_wheel(p_trans)    # Choose a new state
    stochtraj.append(nextX)        # Save the state to the trajectory
    currX = nextX     # Update the current state

fig = plt.figure()
ax = fig.gca()
plt.plot(range(numgen+1),stochtraj,
         color = 'k',
         linestyle = '-',
         linewidth = 2,
         marker = '.',
         markersize = 10)
```

SOLUTION: Calculating Transition Probabilities

The value of $\vec{P}_{t+1}$ for each state can be calculated by multiplying each of the four rows by each of the probabilities $\vec{P}_{t+1}$.

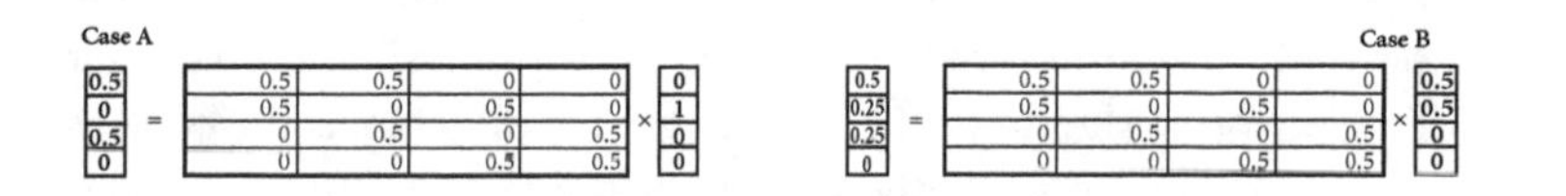

SOLUTION: Dynamics Toward an Absorbing State

The `Tmat` matrix defined above has an absorbing state—state 1. Hence, when the stochastic trajectory reaches state 1, it must remain there (see the first column of the `Tmat` matrix). To illustrate this point, the following figure compares the realized mean from 100 simulations over 75 time points to the theoretical expectation beginning with $\vec{P}_0 = [0, 0, 0, 1]^T$. Keep in mind that the actual mean must be between 1 and 4, so the actual errors are not symmetric. Nonetheless, as is apparent, not only do the mean dynamics agree, but all of the simulations converge to state 1 by the fiftieth time point—and they remain there, permanently.

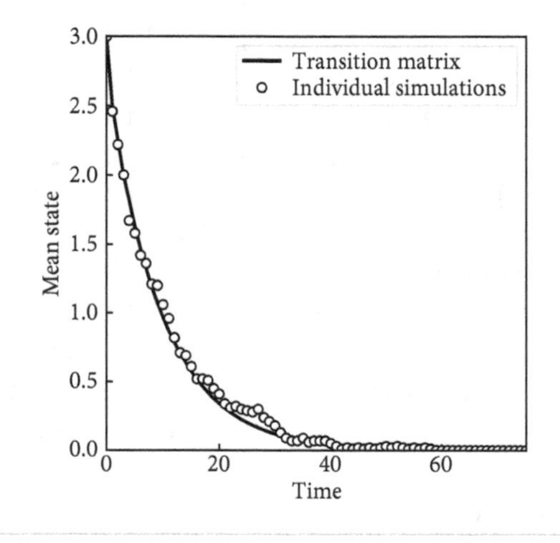

SOLUTION: Explicit Wright-Fisher Model

The key to completing this problem can be broken down into the following steps, as documented in this main loop:

```
for g in range(numgen):
    for i in range(N): # Identify parent of each new individual
        tmpid = np.random.randint(N) # Random parent
        x[g+1, i] = x[g, tmpid] # Offspring type is same as parent
    fracA[g+1] = sum(x[g+1, :] == 1)/N
```

For each generation, the code sweeps through the offspring using the dummy variable `i`. The parent of each offspring is chosen randomly from 1 to N. This is because, in a WF model, the fitness of all parents is equal, so each has the same chance to be the parent of a given offspring. This model does not limit the number of offspring a given parent can have, although the expected number of offspring should be the same for all parents—one. Next, the allele of the offspring should be the same as the parent. Hence, the code assigns the allele a value (1 or 2 here) in the $g+1$ generation for individual i to be the same as the allele value `tmpid` in the g generation for the parent.

SOLUTION: Binomial WF Simulation

The `np.random.binomial` function provides the capability to sample from a binomial distribution. The code snippet below identifies that the Markov process at each generation is equivalent to sampling N times with a probability of selecting type 1 (an A allele) of the fraction of A alleles in the prior generation.

```
# Setup
N = 100 # Number of individuals
numA_0 = N/2 # Fraction that are A
g = 100 # Generations
fracA = np.zeros(g+1)

# Initialize
fracA[0] = numA_0/N

for i in range(1,g+1):
    nextA = np.random.binomial(N,fracA[i-1])
    fracA[i]=nextA/N
# Plot results
plt.plot(np.arange(g+1),fracA,color='k',marker='.',\
         markersize=8)
plt.ylim([0,1])
```

Note that not all simulations will end in an absorbing state, but if they do, there is an equally likely chance that the frequency `fracA` will be either 0 or 1.

Part II

Organismal Behavior and Physiology

Chapter Five

Robust Sensing and Chemotaxis

5.1 TOWARD CHEMOTAXIS IN SINGLE-CELLED ORGANISMS

The goal of this lab is to develop computational tools to simulate an individual moving on a 2D surface either toward or away from a chemical signal. Such movements extend beyond random diffusion to *chemotaxis*, i.e., the directed motion of an organism toward or away from a chemical signal. This laboratory also serves as a bridge from Part I on molecular and cellular biosciences to part II on organismal behavior and physiology. As we make this transition, it is important to keep in mind: bacteria (and archaea) are organisms too!

To get all the way to simulating chemotaxis requires the translation of more than just a few concepts; at minimum, some understanding of how to represent the conversion of extracellular chemical concentrations into an intracellular signal. The next step is to combine such principles with movement dynamics. In essence, the study of chemotaxis already integrates concepts of organismal behavior and physiology, albeit at a microbial scale. Because we are focusing on *E. coli*, the movement will be centered on "runs" and "tumbles." The laboratory also helps to prepare you for independent work: integrating signal transduction and movement into a unified model of chemotaxis (a subject treated in depth in the homework problems).

Notably, chemical concentration gradients may change over time. Hence, this lab covers a set of core techniques that can bridge the gap between stochastic models of random walks and simulations of chemotaxis in single-celled organisms. The key biological concept is that individual bacteria can detect subtle changes in the concentration of chemicals. Some of these chemicals may be beneficial to the bacteria and others may be detrimental. Bacteria, including *E. coli*, have evolved to change their behavior so that they preferentially move toward chemoattractants and away from chemorepellants. In an overly simplistic sense, this process of chemotaxis, i.e., movement driven by chemicals, can be broken down into three core steps. First, bacteria can detect extracellular chemicals via receptor-ligand processes akin to enzyme kinetics. Second, interactions then initiate a signal transduction cascade, mediated by enzyme kinetics. Third, changes in intracellular concentrations of key proteins modulate the behavior of the bacteria, enabling them to have chemotaxis behaviors. As such, the laboratory covers a few essential elements:

- Understanding the basics of signal transduction, from Michaelis-Menten and beyond
- Including time-dependent forcing functions and parameters into numerical integration of coupled ordinary differential equations
- Reviewing probability distributions and their moments
- Building a stochastic spatial model involving diffusion
- Measuring the expected properties of diffusion

These techniques are integrated with both the main text and associated homework. As is typical, these elements are the start of, but not the final word on, investigating how cells integrate information to change and adapt their behavior to local conditions.

5.2 ENZYME KINETICS

Signal transduction describes the process by which cells detect a chemical signal and then transform that external signal into an intracellular signal (Figure 5.1). Usually, this process

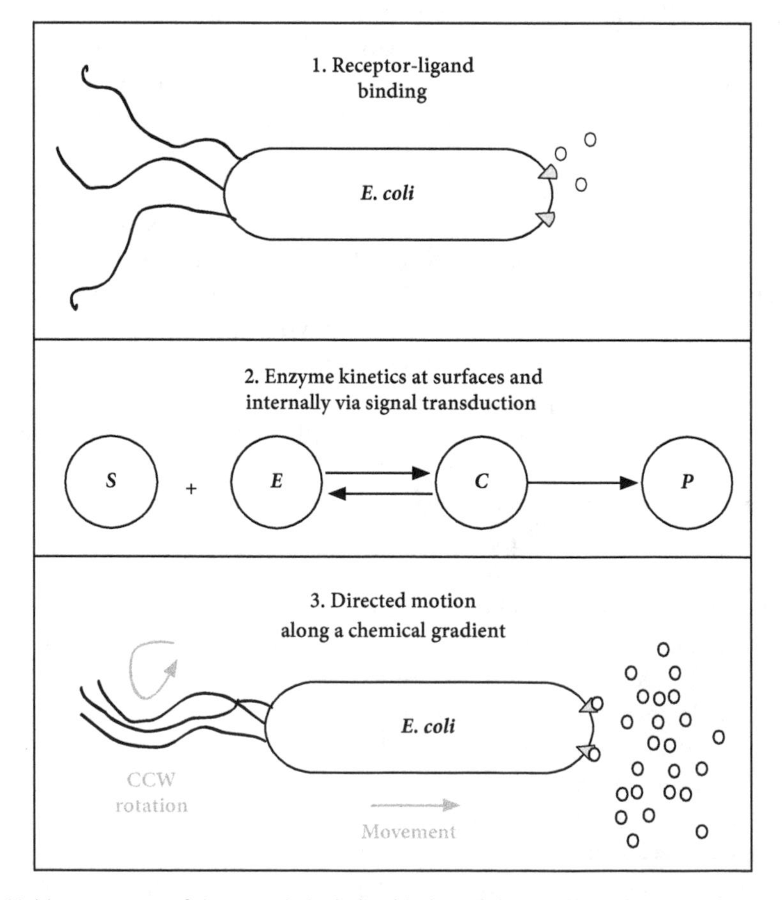

Figure 5.1: Multipart process of chemotaxis, including binding of chemical ligands to receptors, signal transduction, and change in cellular physiology and behavior.

unfolds when a chemical "ligand" binds to a transmembrane receptor. This binding leads to a change in the chemical state of the receptor, in which the intracellular end of the receptor molecule then modifies a diffusible protein. This modification can be a (de)phosphorylation or a (de)methylation event. The core modeling framework that underpins signal transduction is chemical reaction theory. And modeling chemical reactions requires some way of representing perhaps the simplest approximation of changes in chemical concentrations: Michaelis-Menten kinetics.

Consider a chemical reaction in which a substrate S (e.g., a ligand) binds reversibly to an enzyme, E, forming a complex, C. This complex can then generate a product, P. The dynamics of the concentration of this system can be described as

$$\dot{S} = -k_{+}SE + k_{-}C \tag{5.1}$$

$$\dot{E} = -k_{+}SE + (k_{-} + k_{f})C \tag{5.2}$$

$$\dot{C} = k_{+}SE - (k_{-} + k_{f})C \tag{5.3}$$

$$\dot{P} = k_{f}C \tag{5.4}$$

Translating these equations into a computational model is the first challenge problem.

CHALLENGE PROBLEM: Simulating Chemical Reactions

Fill in the following dynamical function to recapitulate the core numerical integration of enzyme kinetics:

```
def mm_dyn(y,t,pars):
    # mm_dyn(t,y,pars):
    # Simulates a full enzyme kinetics model
    # Assign variables
    S, E, C, P = y[0], y[1], y[2], y[3]
    # Core dynamics
    dydt=np.zeros(4)
    dydt[0]=...
    dydt[1]=...
    dydt[2]=...
    dydt[3]=...
    return dydt
```

It is now time to simulate a typical reaction. Consider a ligand (the enzyme) that is present in μM concentrations, with binding rates of 0.1 sec$^{-1}\mu$M^{-1}, unbinding rates of 100 sec^{-1}, and forward reaction rates of 1 sec^{-1}. Use the following code base to simulate and visualize the dynamics:

```
# Parameters - uM and sec units
pars={}
```

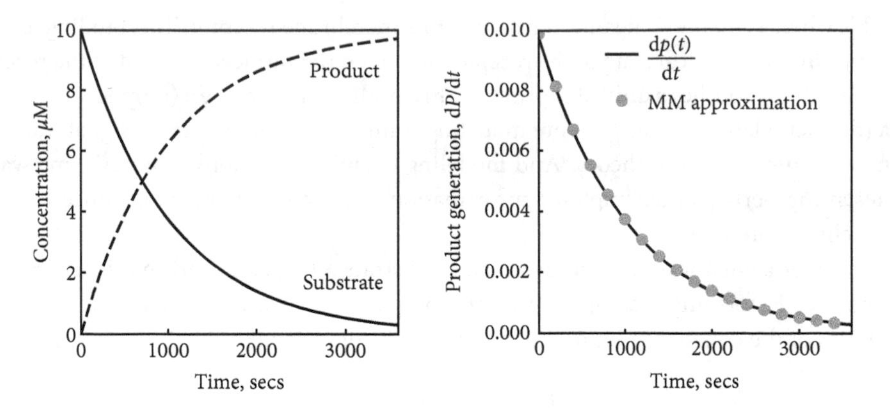

Figure 5.2: Dynamics of full enzyme kinetics, including substrate and product (left) with the MM approximation (right).

```
pars['kplus'] = 0.1
pars['kminus'] = 100;
pars['Km'] = pars['kminus']/pars['kplus']
pars['kf']=1
pars['tf']=3600

# Simulation
t = np.linspace(0,pars['tf'])
y = integrate.odeint(mm_dyn,[10,1,0,0],t,args=(pars,))

# Plot
plt.plot(t,y[:,0], color='k', linestyle='-', linewidth=3)
plt.plot(t,y[:,3], color='k', linestyle='--',linewidth=3)
```

Figure 5.2 shows the resulting depletion of substrate followed by generation of product.

CHALLENGE PROBLEM: Product Dynamics

Compare and contrast the *rate* of product dynamics, i.e., dP/dt, with that anticipated via the Michaelis-Menten (MM) approximation:

$$\frac{dP}{dt} \approx k_f E_0 \frac{S(t)}{k_-/k_+ + S(t)} \tag{5.5}$$

In Figure 5.2 (right), the approximation is compared to that of the full solution. The challenge is to recapitulate this for yourself, i.e., comparing and contrasting the actual product generation rate versus that of the approximation. If you really want an extra challenge, try to identify conditions in which the approximation breaks down.

5.3 TIME-DEPENDENT FUNCTIONS IN DIFFERENTIAL EQUATIONS

Thus far the laboratories have focused on state-dependent changes in differential equations. However, changes may also be time dependent. As such, when modeling environments with external forcing, it is essential to include time-dependent parameters and/or functions. Time-dependent parameters are easily implementable in `integrate.odeint`. Recall the mRNA dynamics from the lab in Chapter 2 given simple high/low production rates and a fixed decay rate. Now consider the following time-dependent ODE for mRNA dynamics:

$$\frac{dm}{dt} = \beta(t) - \alpha m$$

Here the production rate of mRNA, $\beta(t)$, depends on time. Suppose the gene is turned on until $t = 10$ hours. Previously, we would simulate two different ODEs in sequence, one with a β term and one without β. This can be simplified as a single equation:

$$\frac{dm}{dt} = \Theta(10 - t)\beta - \alpha m$$

where we use the Heaviside function, $\Theta(x)$, which is one when $x > 0$ and zero otherwise. A similar equation can be called by `integrate.odeint`:

```
def pulse_production(x,t):

    alphaval = 1 # Decay rate, hrs^-1
    betamax = 5 # Production rate, e.g., nM/hr

    return (t<10)*betamax - alphaval*x
```

where we define the parameter values in the function scope. Note that parameters can be made time dependent when defining the dynamical system.

CHALLENGE PROBLEM: Time-Dependent Input to Dynamical Systems

Include the time-dependent mRNA dynamics and develop your own simulation over a 20-hour period. The resulting dynamics starting with 0 mRNA should look like the left panel:

Next, you can use more exotic functions than simple On and Off switches. Change the code above such that the production rate follows $\beta(t) = (sin(2\pi t) + 1)\beta_{mid}$ with $\beta_{mid} = 5$—this ranges from a minimum of 0 nM/hr to 10 nM/hr. You should now get oscillatory dynamics as in the right panel. How does the time period and the phase (where the peak is located) of the mRNA compare with the production rate?

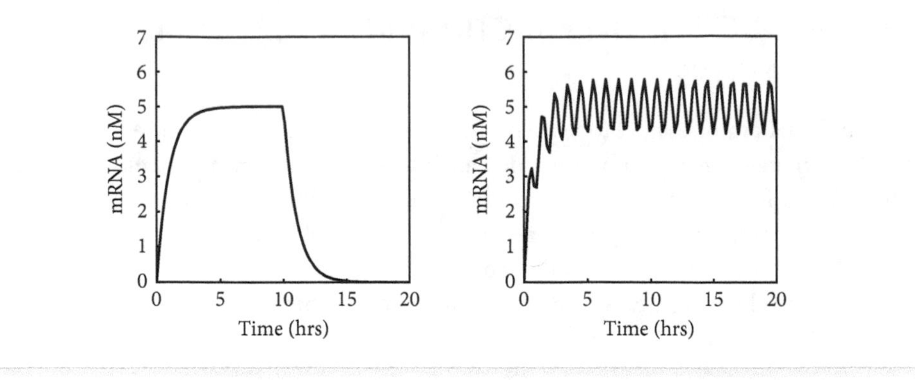

5.4 PROBABILITY DISTRIBUTION REDUX

Previous labs in Chapters 3 and 4 introduced the concept of the difference between an individual stochastic trajectory and the ensemble. The ensemble is a very large collection of trajectories given the same governing parameters. In the case of Markov chains, we found that averaging the properties of many stochastic trajectories, e.g., mean and variance, coincided nearly perfectly with the predictions given the theoretical probability distribution obtained via the transition matrix. Here we make the link between individual samples of a probability distribution and the limiting probability distribution itself. Such a link will be important in building a model of *E. coli* movement.

5.4.1 Binomial distributions

First, suppose we're flipping 15 fair coins (equal chance heads or tails) and want to know the expected number of heads. Recall that the number of heads observed should follow a binomial distribution with $N = 15$ and $p = .5$, with an average number of heads of $7.5 = 15 \times 0.5$. However, the most likely outcome should be either 7 or 8 heads in any individual "simulation." How can we move from the expected probability distribution to a collection of random samples? Below we evaluate three ways.

Theory Use `stats.binom.pmf` to obtain the probability distribution over the discrete support `0:N`.

Sampling using custom code As an alternative, you could sample binomial numbers from this discrete distribution using `randi_wheel` from Chapter 4.

```
import numpy as np
from scipy import stats
# Function
def randi_wheel(fitness):
    tmp_weights = fitness/sum(fitness)
    tmp_order = np.cumsum(tmp_weights)
    x = np.argmax(np.random.uniform()<=tmp_order)
    return x
```

```
# Script
N=15
p=0.5
nvals = np.arange(N+1) # The actual support of the pdf
mypdf = stats.binom.pmf(nvals,N,p)
# Generate one random number using mypdf
rand_bino = nvals[randi_wheel(mypdf)]
```

It is important to use `nvals[randi_wheel(mypdf)]` rather than `randi_wheel(mypdf)`. The reason is that randomly generated numbers from a fixed pdf will return an index. Hence, the first index corresponds to the value of 0 in the binomial distribution, the second index to 1, and so on, such that the $N+1$ index corresponds to the value N, as expected.

Built-in functions The numpy library has a function for sampling binomially distributed random numbers, i.e., `rand_builtin = np.random.binomial(N,p)`.

CHALLENGE PROBLEM: Binomial Distributions

Compare the results of sampling 10^4 different events using a customized code, the appropriate numpy function, and the theoretical distribution for a binomial process, each with $N = 15$ trials and a $p = 0.5$ probability of success.

5.4.2 Gaussian (or normal) distributions

Notice that the binomial distribution is unimodal—it has a distinct peak. There are many distributions with distinct peaks, including the normal (or Gaussian) distribution, which is defined over the entire real line and peaks at the mean value μ. The width of the normal distribution is defined by its variance. The function `np.random.normal` outputs normally distributed random samples. Compare histograms derived from 10^5 samples in the following code:

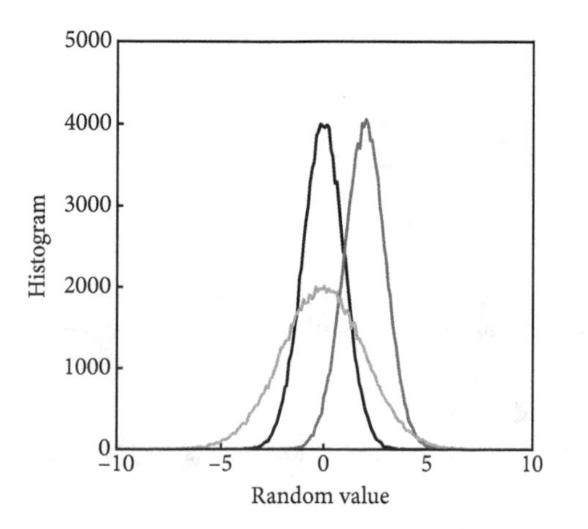

```
norm1 = np.random.normal(size=10**5)
norm2 = np.random.normal(size=10**5)+2
norm3 = 2*np.random.normal(size=10**5)
```

What is the mean and variance for each normal distribution? As should be evident, these three random numbers have means 0, 2, and 0, respectively (as `norm1` is shifted by a factor of 2). However, the shifts do not affect the variance, which is defined as the second central moment, i.e., $E\left((X-\mu)^2\right)$ where X is a random number sampled from a distribution with mean μ. In the above cases, the first two random numbers have variances equal to that of the standard normal distribution—1. However, `norm3` has a variance of 4, because each random number is increased by a constant factor of 2 (whose expected squared deviation is then 4).

5.4.3 Gamma distributions

Another potentially unimodal distribution is the gamma distribution. Recall that the time between events in a Poisson process follows the exponential distribution. But what happens if two events need to take place for a reaction to occur (e.g., multiple bindings leading to the chemical modification of a transcription factor, kinase, or methylase)? How long would it take for $n=2$ sequential Poisson processes to both have events? Let's sample two exponentially distributed vectors and compare the distribution of one to their sums.

```
exp1 = np.random.exponential(3,size=10**5)
exp2 = np.random.exponential(3,size=10**5)
exp12 = exp1+exp2
```

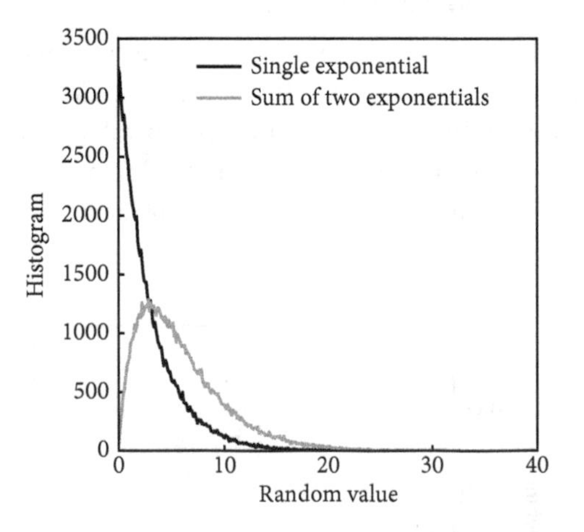

The histogram for a single draw ($n=1$) follows the exponential distribution, but the histogram for the sum of two draws ($n=2$) is unimodal. Compare the mean and variance for each distribution. The results are fairly intuitive; they are both related by a scaling factor of n. What if, instead of the same mean value, the expected times in the multidraw problem scale with $1/n$? Compare the prior results to the following, using this code:

```
exp1 = rexp(10^5,1/3)
exp2a = rexp(10^5,1/1.5)
exp2b = rexp(10^5,1/1.5)
exp12 = exp2a+exp2b
```

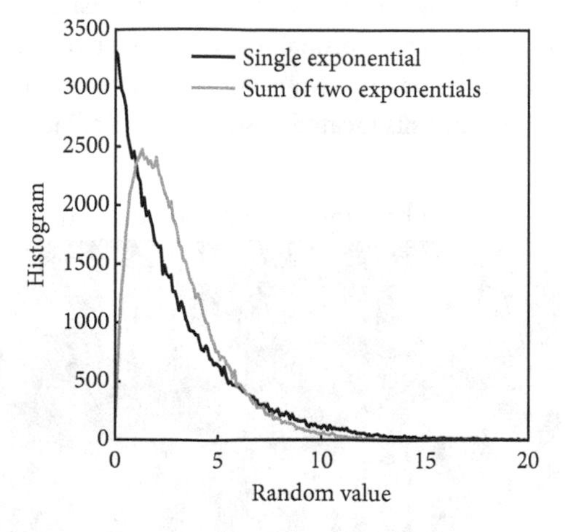

What are the mean and variance for these distributions? Now consider $n = 2$, $n = 3$, and $n = 10$ cases with the same scaling of the rates $\frac{r}{n}$. You will find that the distribution becomes more focused around the mean. This type of dynamics often appears in models of living systems with relatively precise control over timing. For example, models of infectious disease often include multiple intrahost stages of infection. The duration in each state may be exponentially distributed (corresponding to a single rate of progression of the disease). Hence, the overall infected life span is gamma distributed. This concept also applies to viral infections of microbes in which there is a timed process between adsorption, infection, and lysis. More generally, any developmental process with multiple stepwise phases, e.g., the cell cycle or a kinetic cascade, tends to exhibit gamma-like distributions for the total timing of events, making it more likely that events happen close to a given time rather than immediately (as in the exponential distribution). Figure 5.3 illustrates this conceptually.

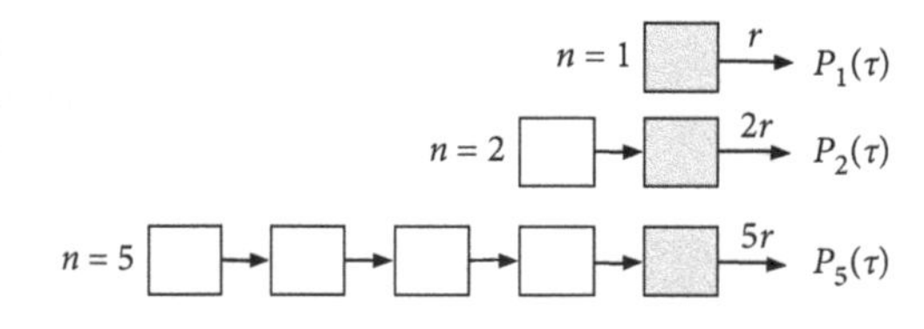

Figure 5.3: Illustration of measuring the distribution of the cumulative time for n independent stochastic processes, each with an individual transition rate of rn. As shown in this section, the average time of $P_n(\tau)$ is exponentially distributed when $n = 1$ and gamma distributed when $n > 1$, i.e., it has a unimodal peak, with ever-narrowing variance as n increases.

This set of calculations raises a question: why does adding up many random numbers each sampled from an exponential yield something unimodal? Some insight can be gained by visualizing the set of event times, t_1 and t_2, such that $t_{tot} = t_1 + t_2$. The puzzle arises when we think about the probability of some combination of event times (t_1, t_2) when each is exponentially distributed with characteristic times τ_1 and τ_2. The joint probability of such event times

$$p(t_1, t_2) = \frac{e^{-t_1/\tau_1} e^{-t_2/\tau_2}}{\tau_1 \tau_2} \tag{5.6}$$

can be seen in the variation of color intensity in the following figure. As you can see, the intensity is brightest at the bottom left, i.e, when both t_1 and t_2 are near zero. However, the white lines denote combinations of event times whose *sum* is equivalent. As you can see, although very short events are the most likely, there are very few combinations of such events (the number of combinations is proportional to the length of the white lines). Similarly, although there are many combinations of events in which the sum is large (long white lines in the upper right), the probability of any such event vanishes exponentially. This argument suggests that there is almost zero probability of having two events happen at once and likely almost zero probability of having two events take a very long time to occur. Hence, the most likely sum of event times is located at some intermediate value!

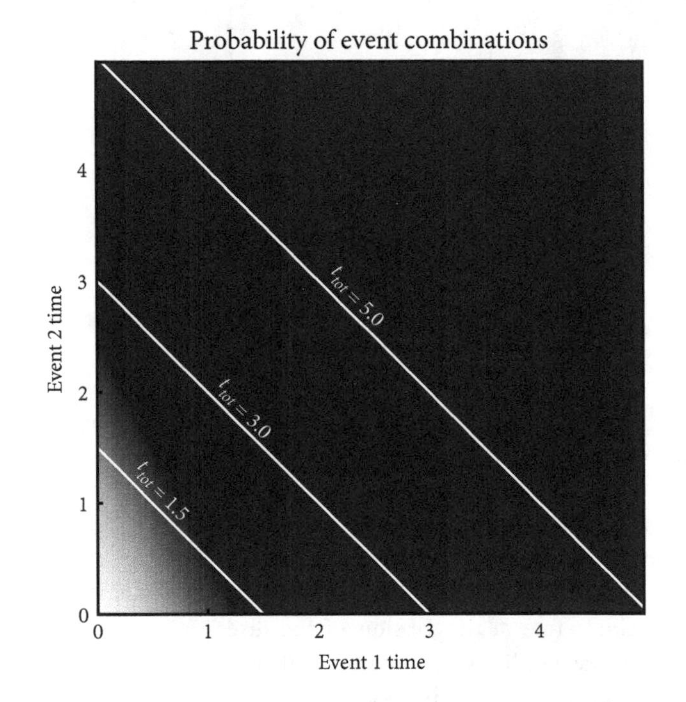

The core elements of this figure were generated through the following script (you can add your own labels and other niceties):

```
# Define the sampling
dt = 0.05
t1 = np.arange(0,5,dt)
t2 = np.arange(0,5,dt)
tau1 = 1
tau2 = 2

# Define the joint probabilities
t1_g, t2_g = np.meshgrid(t1,t2)
p1 = np.exp(-t1/tau1)/tau1
p2 = np.exp(-t2/tau2)/tau2
ptot = np.exp(-t1_g/tau1)*np.exp(-t2_g/tau2)
```

```
# Plot an equal total time to set the axis
# recalling that the implicit dt of the grid is 0.05
fig = plt.figure()
ax = fig.gca()

# Overlay the image
bounds = [0,5,0,5]

# Images index from the top-down, left-right
# Flip the image so that the vertical 0 index is at the origin.
ax.imshow(np.flipud(ptot), extent=bounds)
plt.plot(t1, 5-t1, color='w', linewidth=2)
plt.plot(t1, 3-t1, color='w', linewidth=2)
plt.plot(t1, 1.5-t1, color='w', linewidth=2)
# Finally, set the limits
plt.xlim([0,5])
plt.ylim([0,5])
```

5.5 *E. coli* MOVEMENT

Finally, *E. coli* time! As discussed in the main text, *E. coli* move by running and tumbling. For short distances, "running" involves rotating the flagella in order for the bacterium to move in a straight line at a constant velocity. Tumbling involves rotating the flagella in the opposite direction in order to stop and reorient before running again.

We will develop a model that tracks the cell's (x, y) position over time as it randomly runs and tumbles. Initially, the bacterium runs and tumbles in an environment free from chemical signals. That is, the switch between running and tumbling occurs as a Poisson process with rate λ sec^{-1}. The bacterium runs at a constant speed v between tumbles in units of μm/sec given its current orientation θ. Thus, after a run the x position updates, $x(t+\Delta t) = x(t) + dx = x(t) + v\cos\theta\Delta t$; similarly, the y position updates, $y(t+\Delta t) = y(t) + dy = y(t) + v\sin\theta\Delta t$. Here Δt is the time between runs as sampled from the Poisson process. For the time being, let's assume tumbles occur instantaneously and result in a new randomly selected orientation, θ. We can use the Gillespie algorithm to simulate the stochastic movement dynamics. The following code provides initial values for v and λ, and you are free to modify them to reflect the values for *E. coli* moving through rich media, e.g., $v \approx 30$ μm/sec and $\lambda \approx 1$ sec^{-1}.

```
# Parameters and constraints
tmax = 100 # Seconds
v = 1 # Microns/sec
currangle = np.random.uniform()*2*np.pi
tumblerate = 2 # per sec
```

```
# Initialize the system
currt = 0
currx = 0
curry = 0
X = [[currx, curry]]
t = [0]
ind = 0

# Run and tumble
while currt<tmax:
    ind = ind+1
    dt = -(1/tumblerate)*np.log(np.random.uniform())
    dx = v*np.cos(currangle)*dt # Advance in the x direction
    dy = v*np.sin(currangle)*dt # Advance in the y direction
    currt = currt + dt
    currx = currx + dx
    curry = curry + dy
    t.append(currt)
    X.append([currx,curry])
    currangle = np.random.uniform()*2*np.pi
# Convert X into a numpy array
X = np.array(X)
t = np.array(t)
```

Next, plot the x position versus the y position over the dynamics. Here are three representative curves, which you should aim to reproduce:

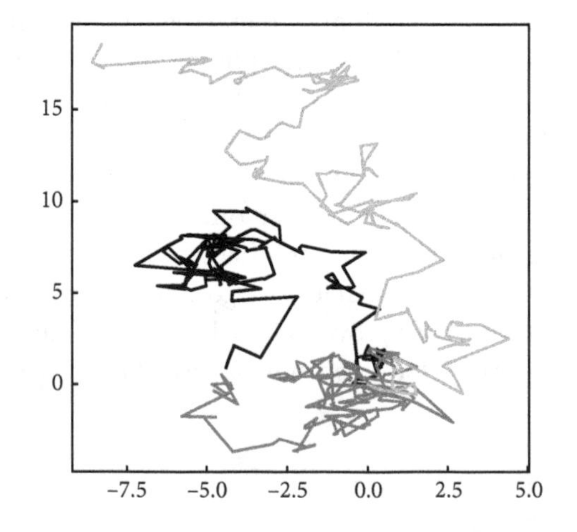

The central point (0,0) marks the origin and the initial condition of all the runs. Notice that these trajectories move in different directions. In these plots, it's hard to address questions related to the timing of movement. For example, how far from the origin is an average bacterium at $t = 50$?

To compare dynamics across an ensemble, we will sample each trajectory at fixed times like we have done in previous labs. However, the process itself involves jumps between new run events. As a consequence, each trajectory need not record precisely the same fixed times. Instead, we need to identify where the bacterium is located at a common set of times. This is an interpolation problem. Consider two points of the x position, x_a and x_b, sampled at times t_a and t_b, respectively, that *bracket* some sample time t_s. That is to say, what is $x(t_s)$ for $t_a \leq t_s \leq t_b$ if we know $x(t_a)$ and $x(t_b)$? We can solve for the x position at $t = t_s$ by linearly interpolating:

$$x(t_s) = x_a + \overbrace{\frac{x_b - x_a}{t_b - t_a}}^{\text{rise over run}} \overbrace{(t_s - t_a)}^{\text{elapsed from prior time}} . \tag{5.7}$$

Graphically, you can see below how doing so interpolates at the sample time $t_s = 1$, which lies between t_a and t_b:

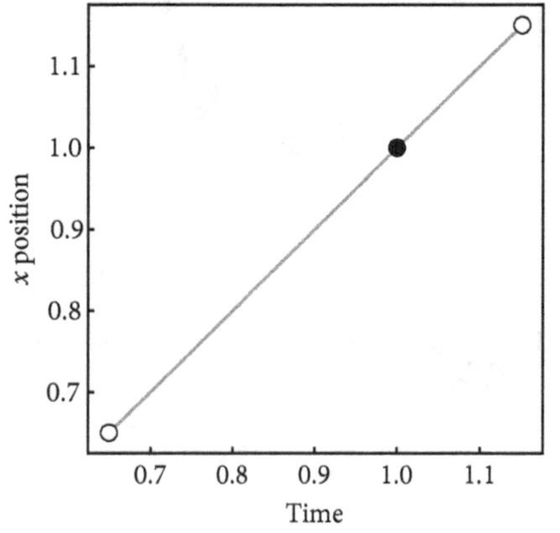

The graph is obtained from the following code:

```
lowdyn = t<1
highdyn = t>1
tlow = t[lowdyn]
ta = tlow[-1]
thigh = t[highdyn]
tb = thigh[0]
xlow = X[highdyn,0]
xa = xlow[-1]
xhigh = X[highdyn,0]
xb = xhigh[0]
deltat = tb-ta
deltax = xb-xa
slope = deltax/deltat
xnew = xa+slope*(1-ta)
```

There is a caveat. Interpolating sparsely can lead to the generation of time series that are not representative of actual paths. Use with caution in selecting an appropriate interpolation time depending on the dynamics of the system at hand.

CHALLENGE PROBLEM: Interpolating Stochastic Trajectories

Adapt the interpolation code above to create a function that converts the stochastic trajectory of positions corresponding to tumbling times to a stochastic trajectory of x and y positions at fixed sample times. (Hint: Loop through all the times you want to record.) Confirm that you properly interpolated by comparing the sampled trajectory to the full trajectory. Here is what the answers should look like if correctly sampled in one dimension:

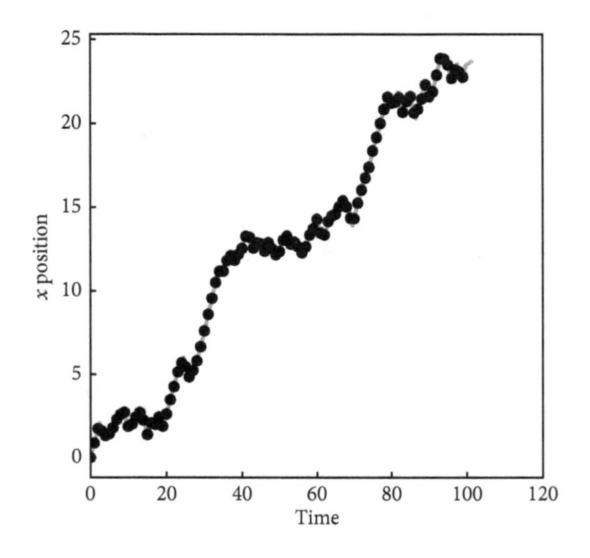

(Hint: There are many ways to accomplish this. The simplest way is to iterate through and interpolate each point. A more efficient way would be to keep track of prior interpolations. However, for the sake of simplicity, we recommend taking the approach that is easier to parse. There is one trick to keep in mind, which is that there is not necessarily a guarantee that the sampling points coincide with the range of the measured points; so your code might crash unless you figure out how to pad the start/finish as needed.)

Run the simulations 1000 times, saving the sampled trajectories. Plot the mean and variance of the x position over time; it should look something like the following.

Although plotting a flat line does not seem instructive, in this case it is meant to emphasize a key point. The expected mean value of the *x position* remains the same as the initial condition, e.g., $x = 0$. In contrast, the variance grows linearly over time. Note that in this interpretation the variance $\sigma^2 = E(x^2) - E(x)^2 = E(x^2)$ is the squared distance. Thus, on average, the *root-mean square distance* a bacterium is from the origin grows as the *square root of time*. This is not an effective means of traversing long distances, but can be efficient over small distances.

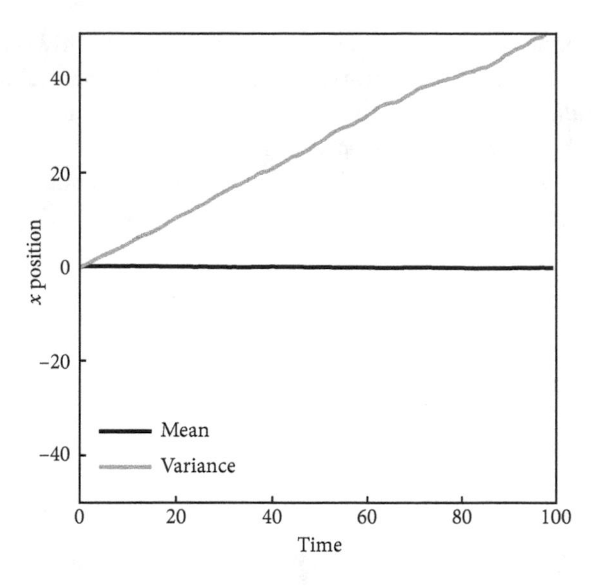

This kind of simulated movement is representative of the physical process of diffusion. This theory is appropriate for particles being thermally bombarded by smaller particles. This is the source of the Brownian motion first identified from pollen particles moving under a microscope. Given Brownian motion, the variance grows linearly over time proportional to the diffusion constant D. Indeed, it is possible to estimate the diffusion constant by fitting a line to the variance dynamics. The diffusion constant has units of length^2 time^{-1}. The dynamics of movement are based on a few parameters: velocity (length time^{-1}), tumbling frequency (time^{-1}), and angle of movement (unitless). Note that via dimensional analysis we expect that the realized diffusion constant should be generated by microscopic mechanisms whose combination has the same units as D. In this case, there is only one way to construct such a value: $D = \frac{\text{velocity}^2}{\text{tumbling frequency}}$. In practice, additional modifications to D are influenced by the typical change in angle in each run, a dimensionless constant, but this is a relatively modest effect compared to the influence of the velocity and tumbling frequency. How does this value compare to your estimate of the slope?

Finally, let's explore how the distribution of x positions evolves over time. Plot the distributions of x positions for $t = 10$, $t = 30$, and $t = 100$.

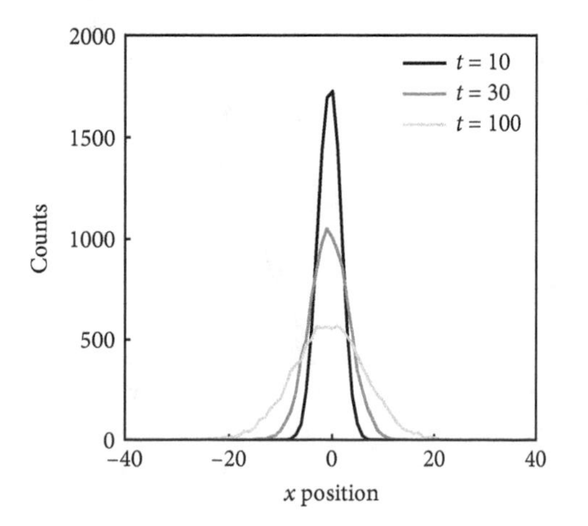

We can visually see the variance growing and the mean staying fairly constant. The distributions are fairly noisy for 1000 trajectories and look similar to normal distributions. Let's compare the cumulative distribution functions (cdfs) of the data and the cdfs of a normal distribution with the same mean and variance as the data. Use the `stats.norm.cdf` command to generate the theoretical cdf. Note: It takes the standard deviation, not the variance, as input.

```
# Calculate statistics
meandata = np.mean(xdist100)
vardata = np.var(xdist100)
sortxpos = np.sort(xdist100)
mycdf = np.arange(1,len(sortxpos)+1)/len(sortxpos)
normtheory = stats.norm.cdf(sortxpos,meandata,np.sqrt(vardata))

# Plot and compare the cdf to theory
fig = plt.figure(figsize=(5,5))
ax = fig.gca()
plt.plot(sortxpos,mycdf,linestyle='',marker='o',
         markerfacecolor='k',markeredgecolor='k')
plt.plot(sortxpos,normtheory,linewidth=2,color=[0.5,0.5,0.5])

# Make the figure look nice
plt.xlabel('x-position',fontsize=20)
plt.ylabel('Probability',fontsize=20)
plt.title(r'$t=100$',fontsize=20)
plt.legend(['Data','Theory'],loc='upper left',frameon=False,fontsize=14)
plt.setp(ax.spines.values(),linewidth=2)
ax.tick_params(labelsize=14,direction='in',width=2)
```

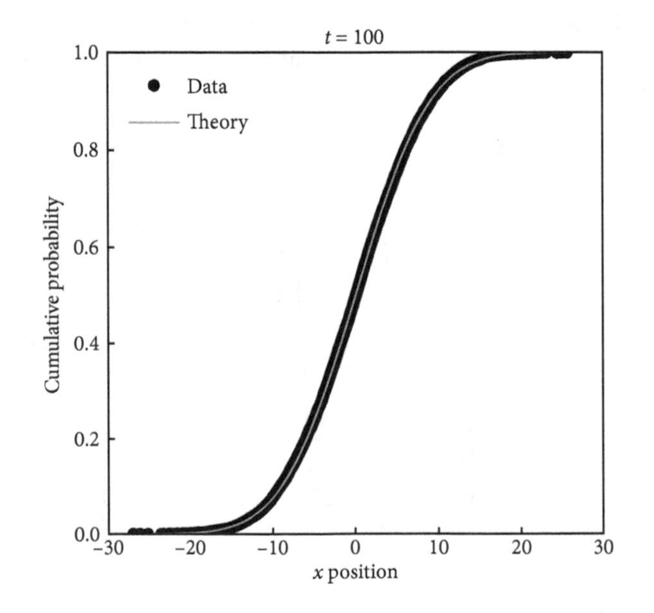

`xdist100` is the data at the focal time of interest. The distribution evolves as a normal distribution over time, as is evident given the overlay of the theoretical cdf and simulation cdf for a normal distribution.

In closing, we note that this baseline model of *E. coli* movement provides a comparison for the actual dynamics described in the textbook and developed further in the homework assignments. The homework enables you to develop a model in which the movement of a particular cell depends on the local chemical state and, even more importantly, on the chemical gradient. Integrating internal states, external stimuli, and motion is a key step to building a quantitative model of an integrative cellular system.

SOLUTIONS TO CHALLENGE PROBLEMS

SOLUTION: Simulating Chemical Reactions

A computational model of enzyme kinetics must ensure that each of the forward and backward reactions are included in the computations as follows:

```
def mm_dyn(y,t,pars):
    # mm_dyn(t,y,pars):
    # Simulates a full enzyme kinetics model
    # Assign variables
    S, E, C, P = y[0], y[1], y[2], y[3]
    # Core dynamics
    dydt=np.zeros(4)
    dydt[0]=-pars['kplus']*S*E + pars['kminus']*C
    dydt[1]=-pars['kplus']*S*E + pars['kminus']*C + pars['kf']*C
    dydt[2]= pars['kplus']*S*E - (pars['kminus']+pars['kf'])*C
    dydt[3]= pars['kf']*C
    return dydt
```

SOLUTION: Product Dynamics

The code to compare and contrast the realized derivative and the expected derivative given the Michaelis-Menten approximation is as follows. Note that there are more sophisticated ways to do this, i.e., by recalculating or storing the instantaneous derivative during simulations. Hence, be mindful that this solution is in its own way an approximation.

```
# Simulate the dynamics
t = np.arange(0,pars['tf'],pars['dt'])
E0 = 1
y = integrate.odeint(mm_dyn,[10,E0,0,0],t,args=(pars,))
```

```
# Compare and contrast the derivatives
dpdt_real = (y[1:,3]-y[:-1,3])/pars['dt']
dpdt_approx = pars['kf']*E0*y[:-1,0]/(pars['Km']+y[:-1,0])

# Plot the comparison
plt.plot(t[:-1],dpdt_real, color='k', linestyle='-', linewidth=3)
# Space out the overlay (for visualization purposes)
plt.plot(t[:-1:200],dpdt_approx[:-1:200], marker='o', linestyle='',
         markerfacecolor=[0.5,0.5,0.5], markeredgecolor=[0.5,0.5,0.5],
         markersize=10)
```

SOLUTION: Time-Dependent Input to Dynamical Systems

A minimal script to simulate mRNA dynamics is as follows:

```
x0 = 0
tmax = 20
t = np.linspace(0,tmax,1000)
mrna = integrate.odeint(pulse_production, x0, t)
plt.plot(t,mrna,linewidth=3)
plt.xlabel('Time',fontsize=20)
plt.ylabel('mRNA',fontsize=20)
```

However, for the more general case, it may be worthwhile to use an anonymous function, e.g.,

```
x0=0
tmax = 20
pars={}
pars['alpha']=1 # hr^-1
pars['beta_mid']=5 # nM/hr
pars['beta'] = lambda t: pars['beta_mid']*(t<10) # nM/hr
t=np.linspace(0,tmax,100)
mrna = integrate.odeint(pulse_production,x0,t,args=(pars,))
plt.plot(t,mrna,linewidth=3,color='k')
plt.xlabel('Time (hr)',fontsize=20)
plt.ylabel('mRNA (nM)',fontsize=20)
```

and then modify the dynamics code to refer to a prespecified function:

```
def pulse_production(x,t,pars):

    return pars['beta'](t) - pars['alpha']*x
```

Note that the dynamics oscillate around the ratio of the average value of $\beta(t)/\alpha$ with a period approaching 1 hour.

SOLUTION: Binomial Distributions

Consider the following script:

```
# Define the parameters and theoretical expectations
N = 15
p = 0.5
xsupp = np.arange(0,N+1)
mypdf = stats.binom.pmf(xsupp,N,p)

# Simulate the ensemble
totruns = 10**5
rand_builtin = np.random.binomial(N,p,totruns)
rand_bino = np.zeros(totruns)
for jj in range(totruns):
    rand_bino[jj]=randi_wheel(mypdf)
```

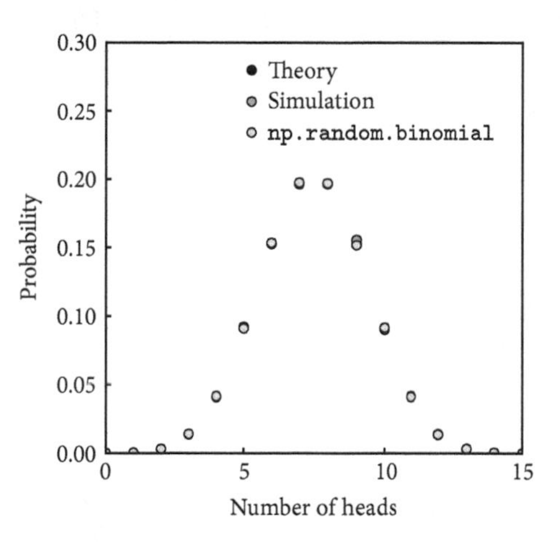

Both the built-in and customized methods of random sampling are equivalent and converge to the same theoretical distribution. Critically, this example helps illustrate the power of the `randi_wheel` function, which enables random sampling from any probability distribution with finite support, i.e., whether parametric or nonparametric. Note that you could modify the customized command to generate different-size matrices of output.

SOLUTION: Interpolating Stochastic Trajectories

The following code interpolates stochastic trajectories. It is but one possible resolution.

```
def interp_2D(t,Pos,samplet):
    # function [samplePos] = interp_2D(t,Pos,samplet)
```

```
# Returns sample values for 2D data in 'pos' given events at time t
# with values in samplePos, which will be resampled
samplePos = np.zeros(shape = (len(samplet),2))
# Earlier sample points will be set to the initial value
tmpi = np.where(samplet <= t[0])[0]
if len(tmpi)>0:
    samplePos[tmpi,0] = Pos[0,0]
    samplePos[tmpi,1] = Pos[0,1]
    indmin = np.where(samplet>t[0])[0][0]
else:
    indmin = 1
# Later points will be set to the final value
tmpi = np.where(samplet>t[-1])[0]
if len(tmpi)>0:
    samplePos[tmpi,0]=Pos[-1,0]
    samplePos[tmpi,1]=Pos[-1,1]
    indmax = np.where(samplet>t[-1])[0][0]
else:
    indmax = len(samplet)

# Resample the trajectory
for jj in range(indmin,indmax):
    # Define the times
    lowdyn = t<=samplet[jj]
    highdyn = t>samplet[jj]
    tlow = t[lowdyn]
    ta = tlow[-1]
    thigh = t[highdyn]
    tb = thigh[0]

    # Define the values
    xlow = Pos[lowdyn,0]
    xa = xlow[-1]
    xhigh = Pos[highdyn,0]
    xb = xhigh[0]
    ylow = Pos[lowdyn,1]
    ya = ylow[-1]
    yhigh = Pos[highdyn,1]
    yb = yhigh[0]

    # Interpolate
    deltat = tb-ta
    deltax = xb-xa
```

```
        slopex = deltax/deltat
        xnew = xa+slopex*(samplet[jj]-ta)
        deltay = yb-ya
        slopey = deltay/deltat
        ynew = ya+slopey*(samplet[jj]-ta)
        samplePos[jj,0]=xnew
        samplePos[jj,1]=ynew
    return samplePos
```

Chapter Six

Nonlinear Dynamics and Signal Processing in Neurons

6.1 COMPUTATIONAL NEUROSCIENCE

The field of computational neuroscience stretches broadly, from neurons to networks to models of cognition. Yet at the core of neuroscience are units of cellular processing: neurons. These neurons are the "nodes" of neural networks. Electrical signals processed at neurons propagate outward along axons and then form the input signals via transmission through dendrites into other neurons. Critically, neurons are not linear transmitters. That is, they do not merely pass on inputs into outputs. Rather, they *process* signals in a nonlinear fashion. This nonlinearity includes the functional ability to filter out some signals and to amplify others. As a consequence, many neurons acting together can respond to signals, store information, recall information, enable short- and long-term memory, and even enable the ability to forget. The origins of nonlinear information processing lie in the fact that neurons are, themselves, dynamical systems. A dynamical system can have the same governing parameters yet be in a different state depending on past history. This also means that the answer to understanding why a particular neuron responds to the same input in a different way might not lie in intrinsic differences in structure, but rather in differences in connectivity, feedback, and its state. That insight underlies the goal of this laboratory: to implement a computational representation of the neuron as a nonlinear dynamical system, following on the seminal work of Hodgkin and Huxley (1952).

The Hodgkin-Huxley (HH) equations denote changes in four components of a neuron: the transmembrane voltage, V, and three gating variables, n, m, and h, associated with the probability that three ionic channels are in the "open" status—potassium, sodium, and a leak (predominantly chloride):

$$\overbrace{C_M \frac{dV}{dt}}^{\text{capacitive current}} = \overbrace{I}^{\text{applied current}} \overbrace{- \bar{g}_K n^4 (V - E_K)}^{\text{potassium current}} \overbrace{- \bar{g}_{Na} m^3 h (V - E_{Na})}^{\text{sodium current}} - \overbrace{\bar{g}_l (V - E_l)}^{\text{leak current}} \tag{6.1}$$

$$\overbrace{\frac{dn}{dt}}^{\text{potassium channels}} = \overbrace{\alpha_n (1 - n)}^{\text{on}} - \overbrace{\beta_n n}^{\text{off}} \tag{6.2}$$

$$\overbrace{\frac{dm}{dt}}^{\text{sodium channels}} = \overbrace{\alpha_m(1-m)}^{\text{on}} - \overbrace{\beta_m m}^{\text{off}} \tag{6.3}$$

$$\overbrace{\frac{dh}{dt}}^{\text{leak channels}} = \overbrace{\alpha_h(1-h)}^{\text{on}} - \overbrace{\beta_h h}^{\text{off}} \tag{6.4}$$

Figure 6.1 provides a schematic of the process. The details and derivation of the HH equations are found in the textbook.

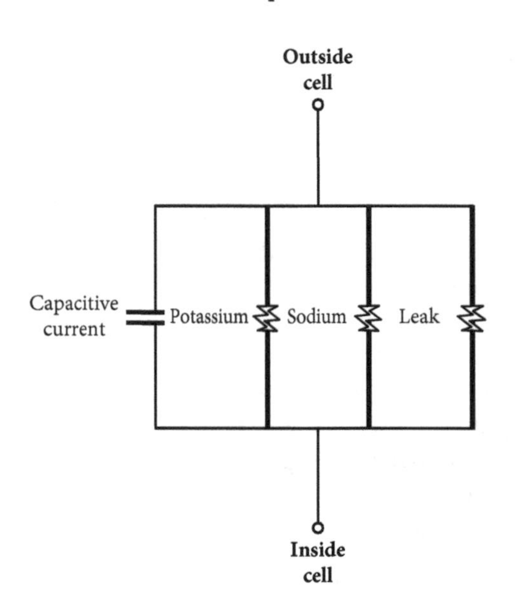

Figure 6.1: Representation of the transfer of ions across a cell membrane via ion channels and via a capacitive current, reflecting the ability of the membrane to hold a charge difference.

Briefly, the phrase above each term in the equations helps to explain the context. The dynamics arise from three physical effects: (i) Ohm's law, i.e., that a voltage is equal to the current multiplied by the resistance (or the inverse of capacitance); (ii) Kirchhoff's rule, i.e., that currents are conserved; (iii) the opening and closing of ionic channels, which are voltage dependent:

$$\alpha_n(V) = 0.01\frac{10-V}{e^{\frac{10-V}{10}}-1} \tag{6.5}$$

$$\beta_n(v) = 0.125e^{-V/80} \tag{6.6}$$

$$\alpha_m(V) = 0.1\frac{25-V}{e^{\frac{25-V}{10}}-1} \tag{6.7}$$

$$\beta(m)(V) = 4e^{-V/18} \tag{6.8}$$

$$\alpha_h(V) = 0.07e^{-V/20} \tag{6.9}$$

$$\beta_h(V) = \frac{1}{e^{\frac{30-V}{10}}+1} \tag{6.10}$$

Throughout these equations, voltage is measured in units of millivolts (mV).

These equations reflect a tour de force of experiments and modeling—but they are also opaque, at least on first encounter. Indeed, what is possible for a neuron to do, in a dynamic sense, may not appear obvious by simply looking at (or admiring) these equations. But that's okay. These equations are not meant to be admired. Instead, they represent a biophysical representation of how a neuron responds to stimuli, and they are meant to be used, explored, and compared to experiments. Hence, the goal of this lab is to simulate neuronal dynamics and to identify some of the key dynamical features of this excitatory system. By doing so, the lab will provide a guide to modulate the Hodgkin-Huxley model, see how it works in practice, and gain intuition for concepts including (i) subcritical inhibition; (ii) supercritical firing; and (iii) refractory behavior. Unlike prior laboratories, simply writing in the code takes time, so it is recommended that some components of this lab be loaded prior to starting. Some of these preloaded codes have missing components; that is intentional to give you an opportunity to fill in the gaps. Even if you have the code in hand, please be sure to read through each line. There is subtlety here, and editing and modifying the code will help build your understanding of computational neuroscience in a way that merely hearing about what neurons can do never will.

6.2 THE HODGKIN-HUXLEY MODEL

In this section, you will "experiment" with the HH model. In doing so, you will have a chance to build your intuition for the mechanisms by which a neuron can be "excited," how it can enter a "refractory" mode, and other consequences of the nonlinear interactions and time scales in the model.

6.2.1 Basic HH model

Recall that the Hodgkin-Huxley model can be written as

$$
\begin{aligned}
C\dot{V} &= I - \bar{g}_K n^4 (V - E_K) - \bar{g}_{Na} m^3 h (V - E_{Na}) - \bar{g}_l (V - E_l) \\
\dot{n} &= \alpha_n(V)(1-n) - \beta_n(V) n \\
\dot{m} &= \alpha_m(V)(1-m) - \beta_m(V) m \\
\dot{h} &= \alpha_h(V)(1-h) - \beta_h(V) h
\end{aligned}
$$

where the changes in the gating variables are modulated by On and Off rates that are themselves voltage dependent (see equations above). The parameters are defined in Hodgkin and Huxley's original paper and already embedded in the codes.

This laboratory leverages a predeveloped HH simulator. The code in Python has four components:

- `master_hh.py`—the master script for running the HH model
- `impulse_t.py`—a function that specifies customized current impulses
- `model_hh.py`—the dynamical system that calculates the change of each of the variables with time
- `plot_hh.py`—an output program that encapsulates the dynamics in a series of preformed panels, using the style and axis scales of Eugene Izhikevich (2007).

You are strongly encouraged to type in the model and master files, as doing so will help you think about the consequences of the particular relationships. The complete file set is available for download on the laboratory home page, but if you choose to download the files, you are again strongly encouraged to slowly read them, i.e., over a 5–10 minute period.

Note that the formulation of the HH problem presents a challenge. The gating dynamics include On and Off rates with forms like $f(V)/g(V)$, where $g(V) = 0$ for some finite voltage V_s. This would seem to imply a singularity, i.e., a divergence in the rate. But $f(V)$ also goes to 0 as $V \to V_s$, such that the ratio has a well-defined limit. Numerical solvers do not necessarily know that. Hence, for some solvers, the simulations of the HH model crash or do not behave correctly, e.g., giving spurious or nonrepeatable results. Hence, there are also two alternative files that you can/should use instead in the event of abnormal behavior. Read through this alternative code set and try to figure out what it does to avoid this "singular" problem:

- `master_alt_hh.py`—the master script for running the HH model and avoiding divide-by-0 errors
- `model_alt_hh.py`—the dynamical system that calculates the change of each of the variables with time, while avoiding divide-by-0 errors

The solution to avoiding divide-by-0 errors is described in the technical appendix to this lab.

6.2.2 The key master and dynamics codes

Here is the master code that sets parameters and runs the model. You will note that a few pieces are missing.

```
# Master_hh
# Script for HH models
#
# Simulates the HH model using parameters embedded in pars
# A stimulus current, and a time range over which the simulation should run

pars={}
# Parameters, including On/Off functions
pars['gKbar'] = 36     # mS/cm^2
pars['gNabar'] = 120   # mS/cm^2
pars['gL'] = 0.3       # mS/cm^2
pars['EK'] = -12        # mV
pars['ENa'] = 120      # mV
pars['EL'] = 10.6      # mV
pars['C'] = 1          # muF/cm^2
pars['alphan'] = lambda V: 0.01*(10-V)/(np.exp(1-V/10)-1)
pars['betan'] = lambda V: 0.125*np.exp(-V/80)
pars['alpham'] = lambda V: 0.1*(25-V)/(np.exp(2.5-V/10)-1)
pars['betam'] = lambda V: 4*np.exp(-V/18)
pars['alphah'] = lambda V: 0.07*np.exp(-V/20)
pars['betah'] = lambda V: (np.exp(3-V/10)+1)**-1

# Initial conditions
pars['V0'] = 0
pars['n0'] = 0
pars['m0'] = 0
pars['h0'] = 0

# Run the model
t0 = 0
tf = 20
tstep = 0.02
t = np.arange(t0,tf+tstep,tstep)
```

```
y0 = np.array([pars['V0'], pars['n0'], pars['m0'], pars['h0']])
y = integrate.odeint(model_hh,y0,t,args=(pars,))

# Store the results
# FILL IN WHEREVER YOU SEE ...
dyn = {}
dyn['t'] = t
dyn['V'] = y[:,0]
dyn['n'] = y[:,1]
dyn['m'] = y[:,2]
dyn['h'] = y[:,3]
dyn['gK'] = pars['gKbar']*...
dyn['gNa'] = pars['gNabar']*...
dyn['gL'] = pars['gL']*...
dyn['IK'] = ...
dyn['INa'] = ...
dyn['IL'] =...
dyn['appliedI'] = impulse_t(dyn['t'])
```

CHALLENGE PROBLEM: Completing the HH Setup

Download the HH code package and then fill in the missing pieces of the `master` code wherever you see an ellipsis (...).

CHALLENGE PROBLEM: Completing the HH Equations

Download the HH code package and then fill in the missing pieces of the `model` code wherever you see an ellipsis (...).

```
# model_hh
def model_hh(y,t,pars):
    # function dydt = model_hh(y,t,pars)
    # This simulates the HH model using parameters embedded in pars
    # A stimulus current in pars['Idrive']

    # Variables
    V = y[0]
    n = y[1]
    m = y[2]
    h = y[3]
```

```
    # Impulses
    I = impulse_t(t) # Specified in a function

    # Dynamics
    Vdot = (1/pars['C'])*(I-...) # Fill in
    ndot = ...
    mdot = ...
    hdot = ...

    # Fill in
    dydt = np.array([Vdot, ndot, mdot, hdot])
    return dydt
```

The script and function used to numerically integrate the Hodgkin-Huxley model also requires the application of an applied current, $I(t)$. Chapter 5 demonstrated how to include time-dependent parameters in the numerical integration of ODEs. Here you should do the same, and consider saving different versions of applied current functions for use and reuse.

CHALLENGE PROBLEM: Time-Dependent Impulses

Develop an impulse function in which the applied current is

$$I(t) = 2 \qquad 2 \leq t \leq 2.5$$

$$I(t) = 25 \qquad 10 \leq t \leq 10.5$$

$$I(t) = 0 \qquad \text{otherwise}$$

all in units of $\mu A/cm^2$. In doing so, modify the following impulse code that enables time-specific additions of current.

```
# impulse_t
def impulse_t(t):
    # function I = impulse_t(t)
    # Specifies the applied time-varying current
    # Works if t is a single value or many values

    if isinstance(t,float):
        if t>2 and t<2.5:
            I=...
        elif t>10 and t<10.5:
```

```
            I=...
        else:
            I=...
    else:
        I=np.zeros(len(t))
        for i in range(len(t)):
            if t[i]>2 and t[i]<2.5:
                I[i] = ...
            elif t[i]>10 and t[i]<10.5:
                I[i] = ...
            else:
                I[i]=...
    return I
```

6.2.3 A digression on plotting multiple axes

The additional codes include plotting capabilities. These plotting codes are essential, but they are also stylized given that they were developed for a particular, aesthetic output, but not necessarily for straightforward reading. There are good ideas here on how to stack axes that should be read and reviewed with care. In addition, the technical appendix includes information on aesthetic choices.

CHALLENGE PROBLEM: Customizing Axes in a Single Figure

Plot the voltage-sensitive gating functions as stacked plots, with a common shared x axis and individual y axes. In Python, you can create both a figure handle and an array of axes with the `plt.subplot` command. When you give this command, you have the option of resizing the figure to make room for multiple axes as in the following:

```
fig, axes = plt.subplots(numRows,numCols,figsize = (figWidth,figHeight))
```

In addition, you can access the individual axes from the array:

```
ax0=axes[0]
```

You can also eliminate tick labels (useful when axes share the same x values), as follows:

```
ax0.set_xticklabels([])
```

If your code is working, the result should look like the images below.

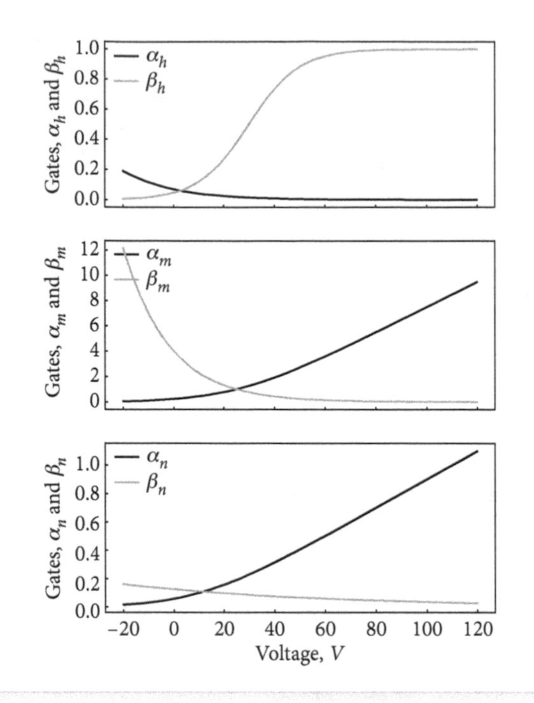

6.3 FIRING WITHOUT A CURRENT

Once you have the code entered or downloaded, you should be able to run the model by compiling `master_hh.py`. The resulting plot takes up a lot of space, making it difficult to view in the console. If all goes well, you should get the result shown in Figure 6.2 (left). As is apparent, the neuron seems to begin to fire despite the fact that no current is applied. In addition, the neuron seems refractory with respect to the larger applied current at 10 ms. In contrast, the same model using the same parameters yields the "classic" neuronal excitation dynamics, as seen in Figure 6.2 (right). The plotting code does include the detailed annotations, as did Izhikevich; however, you should already notice a number of differences.

Thought Problem for Reflection

Identify the difference between the initial run and that shown in Figure 6.2. What is the biological meaning of the difference and why does the neuron "fire" before a signal arrives? Can you identify a "fix" to the Python model that would make it behave "correctly"?

Spoiler alert next . . .

Figure 6.3 illustrates that the model developed here (left) and the HH model (right) are exactly the same. The applied currents are more or less the same. The initial voltage is

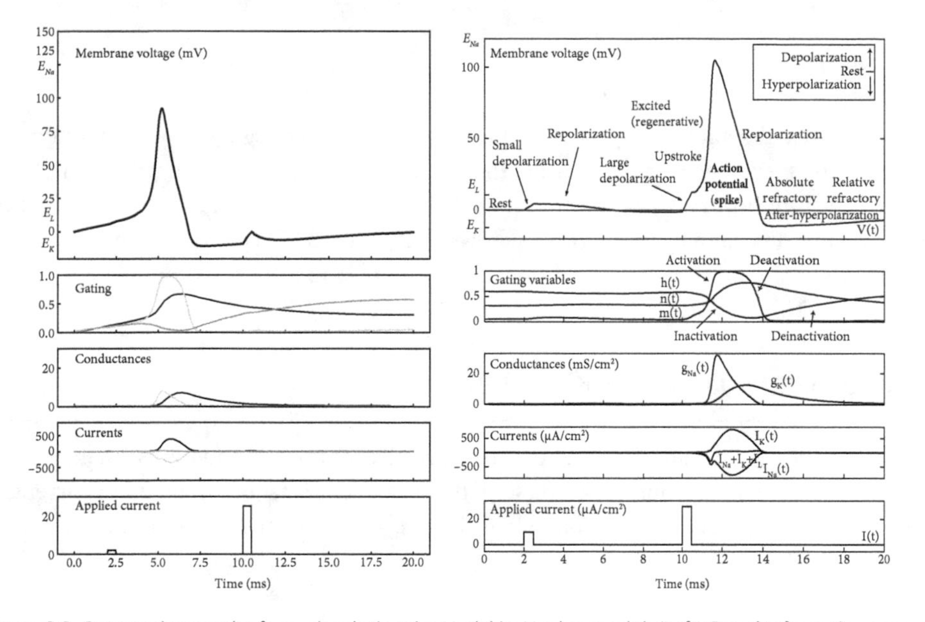

Figure 6.2: Contrasting results from simulating the Hodgkin-Huxley model. (Left) Results from the presupplied R Python code. (Right) Results from Figure 2.15 of Izhikevich (2007).

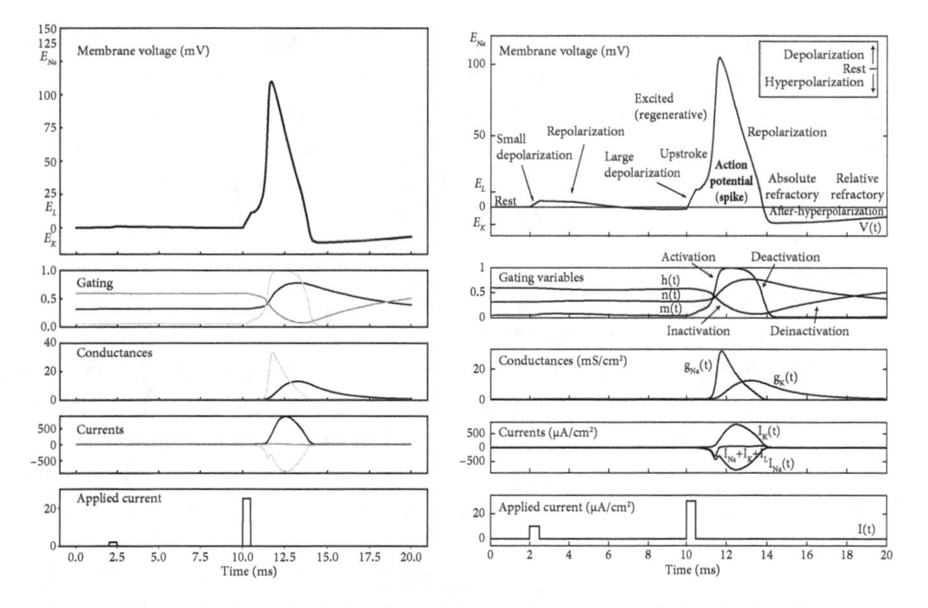

Figure 6.3: Agreeing results from simulating the Hodgkin-Huxley model. (Left) Results from the presupplied Python code. (Right) Results from Figure 2.15 of Izhikevich (2007).

the same. However, previously, the initial gating states were all set to 0, i.e., they were all inactive. This is not the equilibrium condition of the model. Note that at equilibrium

$$n^* = \alpha_n(V^*)/(\alpha_n(V^*) + \beta_n(V^*))$$
$$m^* = \alpha_m(V^*)/(\alpha_m(V^*) + \beta_m(V^*))$$

$$h^* = \alpha_h(V^*)/(\alpha_h(V^*) + \beta_h(V^*))$$

At these particular values, the system is at rest. If not, the gates begin to open, conductances increase, the membrane depolarizes and then repolarizes, and the system will appear to fire even if no stimulus or a weak stimulus is applied. If one uses the equilibrium values as the initial conditions, then the results are as shown in Figure 6.3. As is evident, the neuron does not fire given the first applied current but does so in response to the second, larger applied current at 10 ms.

CHALLENGE PROBLEM: Filtering and Excitation

Identify the equilibrium voltage and gating variables for the HH equations. Then, using the same impulse function as before, see if the system can filter out small input currents and react in an excitatory fashion to large inputs, as demonstrated in Figure 6.3.

6.4 NEURON DYNAMICS: THRESHOLDS IN MAGNITUDE AND TIME

Now that you have a functioning HH model, it is time to explore. The focus of these last two challenge problems is to modify the input $I(t)$ in different ways to identify how a neuron can exhibit threshold behavior given variation in the magnitude of an incoming pulse, as well as the spacing between pulses. However, please do not feel constrained by only these two challenge problems. There is no way to fully cover the scope of neuronal dynamics in this laboratory, and self-guided exploration of the model is essential.

CHALLENGE PROBLEM: Thresholds for Firing

Modify the code to identify the minimum current required for firing. Such current can depend on the duration of the pulse. Here identify the minimum applied current over a period of 0.5 msec required to fire the neuron. You can cross-check your answer by verifying that applying a smaller current (e.g., 1 μ A/cm^2) does not lead to a firing event. You should find a result as in Figure 6.3 where the bottom $I(t)$ panel has been removed.

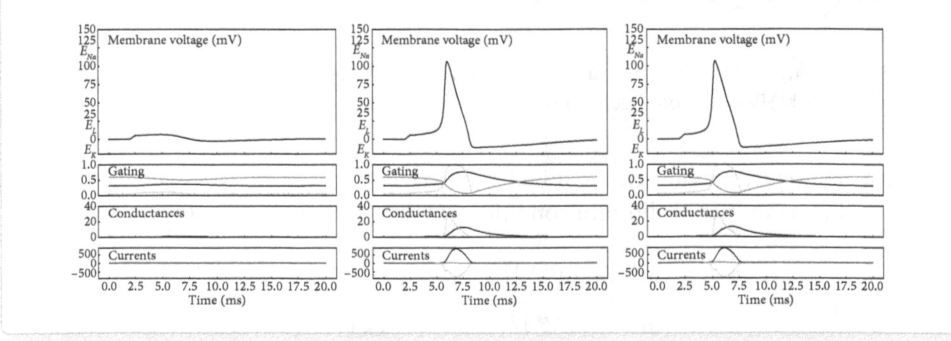

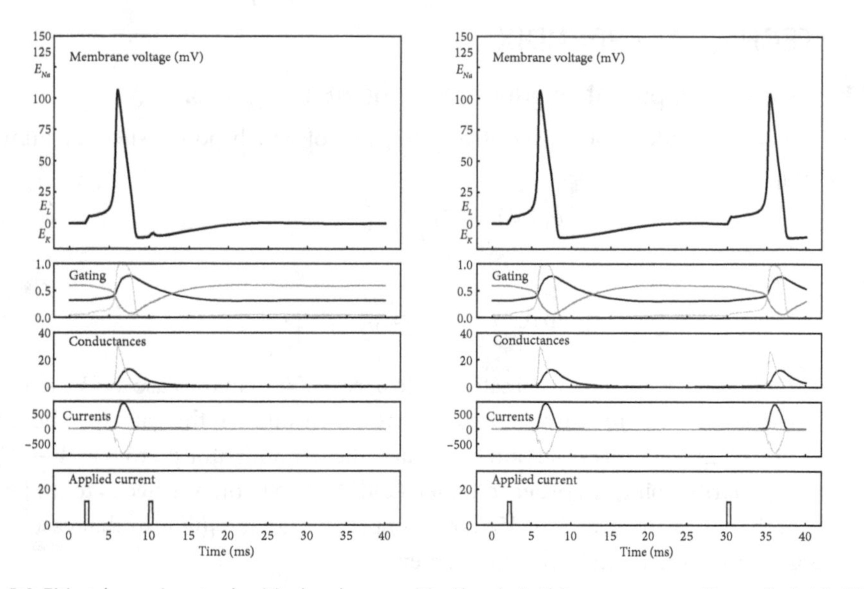

Figure 6.4: Firing dynamics at subcritical and supercritical levels. In this case, a current is applied at 2–2.5 ms with the magnitude differing by only 1.

Next, modify the impulses to explore the effects of the timing interval between the applied current and the resulting firing of neurons. An example is provided in Figure 6.4. In it, you will see that there is a supercritical applied current that drives an initial action potential. In the two scenarios, the second applied current pulse has markedly different effects on the next action potential.

BONUS CHALLENGE PROBLEM: Refractory Periods

What is the minimum interval required to induce a second action potential spike in the neuron after the first applied current? Note that the answer to this problem is not provided here; it is ideally suited for in-depth exploration.

There are many other ways to explore neuronal dynamics and excitation. Here are just a few possibilities:

- Neuronal dynamics exhibit many features, including hyperexcitability. Can you identify a period after the initial pulse where even *less* current is required to generate an action potential spike?
- Can you characterize the time scales of relaxation of the activation and inactivation gates? How do they differ as a function of voltage?
- How do the period and peak height of the action potential vary with the intensity of the applied current? Is the relationship linear or nonlinear?
- Explore the effect of additional impulses of different duration and magnitudes while keeping the product of the current times the time the same. Are all "pulses" equal? Can a long, weak pulse generate a signal? Can a long, but very strong pulse generate a signal?

6.5 TECHNICAL APPENDIX

6.5.1 Avoiding apparent singularities in the HH integration

The HH model includes voltage-sensitive gates, two of which pose issues for numerical integration:

$$\alpha_n(V) = 0.01\frac{10 - V}{e^{(10-V)/10} - 1} \tag{6.11}$$

and

$$\alpha_m(V) = 0.1\frac{25 - V}{e^{(25-V)/10} - 1}. \tag{6.12}$$

The issue for both is that the denominators go to 0, e.g., $V = 10$ in the case $\alpha_n(V)$ and $V = 25$ in the case of $\alpha_m(V)$. Yet in neither case does the actual value of the gate-specific On rate diverge. Instead, in both cases the numerator and the denominator go to 0 as $V \to V_s$ (the apparent singularity point). In order to understand the behavior, one needs to understand the limit of these functions, as $V \to V_s$. For example, consider, $\alpha_n(V)$. In that case, we can use l'Hôpital's rule, i.e., if $q(x) = f(x)/g(x)$, then

$$\lim_{x \to x_0} q(x) = \frac{f'(x)|_{x=x_0}}{g'(x)|_{x=x_0}} \tag{6.13}$$

where $f'(x)$ and $g'(x)$ denote derivatives with respect to x. In which case

$$\lim_{V \to 10} = 0.01\frac{-1}{-1/10 \times e^{(10-V)/10}|_{V=10}}, \tag{6.14}$$

which means $\alpha_n(10) = 0.1$. Similarly, then $\alpha_m(25) = 1$.

It would seem that one could run a numerical integrator but always utilize a replacement if the integration ever passed precisely on the value of the apparent singularity. Not all numerical integrators work this way, i.e., replacing just one point rather than a continuous set of points poses a problem for schemes. Instead, one way is to use a piecewise smooth version of the functions. The following code replaces the original version shown in this lab; it is a bit less intuitive, but it is also more reliable.

```
# Define the expected voltage limits
pars['Vmin'] = -100
pars['Vmax'] = 200
pars['dV'] = 0.1
V = np.arange(pars['Vmin'],pars['Vmax']+pars['dV'],pars['dV'])
pars['V'] = V

# Define the original functions, with piecewise smooth functions
pars['alphan'] = 0.01*(10-V)/(np.exp(1-V/10)-1)
pars['alpham'] = 0.1*(25-V)/(np.exp(2.5-V/10)-1)
# Anonymous functions
pars['alphah'] = lambda V: 0.07*np.exp(-V/20)
pars['betan'] = lambda V: 0.125*np.exp(-V/80)
pars['betam'] = lambda V: 4*np.exp(-V/18)
pars['betah'] = lambda V: (np.exp(3-V/10)+1)**-1
```

```
# Assign the l'Hopital limit to those values where
# Python thinks the function is 'Not a Number' (NAN)
tmpi = np.where(np.isnan(pars['alphan']))[0]
pars['alphan'][tmpi]=0.1
tmpi = np.where(np.isnan(pars['alpham']))[0]
pars['alpham'][tmpi]=1
```

With the script in hand, you should notice that the function invoked by `integrate.odeint` must also change. The derivatives must now utilize the interpolated values of `pars['alphan']` and `pars['alpham']`. The code below is the core of the function `model_hh` in which the following steps are taken. First, the index of the interpolated array of α_n and α_m is found. Next, the values of α_n and α_m are found by assuming that these rates can be linearly interpolated on a tight grid. Finally, the values are utilized for the exact same set of HH equations as before.

```
# model_hh
def model_hh(y,t,pars):
    # function dydt = model_hh(y,t,pars)
    # This simulates the HH model using parameters embedded in pars
    # A stimulus current in pars['Idrive']
    ...

    # Fixes the issue with the singularity in the alphan/alpham
    # The intuition here is to find the right index for
    # the current V value, and then interpolate to make it piecewise
    # smooth. This method also avoids using ''where'' and ''if'' commands
    # to ensure reasonable speed.
    vmin_ind = 1+(V-pars['Vmin'])/pars['dV']
    tmpi = int(np.floor(vmin_ind))
    dev_ind = vmin_ind - tmpi
    alphan = pars['alphan'][tmpi] + (pars['alphan'][tmpi+1]-pars['alphan'][tmpi])*dev_ind
    alpham = pars['alpham'][tmpi] + (pars['alpham'][tmpi+1]-pars['alpham'][tmpi])*dev_ind

    # Dynamics
    Vdot = (1/pars['C'])*(I-pars['gKbar']*n**4*(V-pars['EK'])-pars['gNabar']*m**3*h*
                         (V-pars['ENa'])-pars['gL']*(V-pars['EL']))
    ndot = alphan*(1-n)-pars['betan'](V)*n
    mdot = alpham*(1-m)-pars['betam'](V)*m
    hdot = pars['alphah'](V)*(1-h)-pars['betah'](V)*h

    # Fill in
    dydt = np.array([Vdot, ndot, mdot, hdot])
    return dydt
```

6.5.2 Plotting—it takes time

Plotting—in whatever language—takes time and effort. The beginning coder is encouraged to hew close to these basic guidelines, and perhaps consult one of Edward Tufte's many books on the quantitative display of visual information for inspiration:

- Every figure with data should have its own scripted figure file; i.e., if the figure was worth making once, it may be worth making twice, and you should be able to do the exact same thing again!
- The ratio of data to ink should be high. That is to say, white space should be respected. As but one example, if you have a "bar" chart, consider replacing the tops

of ink-filled bars with single points. Same data, less ink. Apply principle repeatedly until satisfied.

- Ranges and axes should be used to help guide the reader; i.e., judicious use of axes, ranges, and scaling (log or linear) can help guide a reader to improved interpretation.
- Use color. Papers are generally read online. (for those reading this book, we concede that the constraints of the publication process into a physical book mean that the bulk of the figures in these laboratories are in black-and-white. Feel free not to be constrained by such a limitation in your own work.)
- Legends, labels, and units: Use them.

Keeping this preamble in mind as you proceed, we turn to the following command, which generates a plot and returns a handle based on the dynamics embedded in `dyns` and the parameters in `pars` resulting from an HH simulation.

```
def plot_hh(dyn,pars,savePDF=False,figName=''):
    # function ploth = plot_hh(dyn, pars)

    # Takes an HH output structure in dyn and plots dynamics of voltage,
    # gating variables, conductance currents, and applied current

    fig, axes = plt.subplots(5,1,figsize = (9,12),gridspec_kw = {'height_ratios':[4,1,1,1,1]})



    # Main data goes here

    # Applied current on bottom panel
    ax4 = axes[4]
    ax4.plot(dyn['t'],dyn['appliedI'],color='k', linestyle='-',linewidth=2)
    ax4.set_xlabel('Time (ms)',fontsize=12)
    ax4.text(0,23,'Applied current',fontsize=20)
    ax4.set_ylim([0,30])
    plt.setp(ax4.spines.values(),linewidth=2)
    ax4.tick_params(labelsize = 20,width=2,direction='in')

    # Next up is currents
    ax3 = axes[3]
    ax3.plot(dyn['t'],dyn['IK'],color='k')
    ax3.plot(dyn['t'],dyn['INa'],color='r')
    ax3.plot(dyn['t'],dyn['IL'],color='b')
    ax3.text(0,300,'Currents',fontsize=20)
    ax3.set_ylim([-900,900])
    plt.setp(ax3.spines.values(),linewidth=2)
    ax3.tick_params(labelsize = 20,width=2,direction='in')
    ax3.set_xticklabels([])

    # Conductance
    ax2 = axes[2]
    ax2.plot(dyn['t'],dyn['gK'],color='k')
    ax2.plot(dyn['t'],dyn['gNa'],color='r')
```

```
ax2.plot(dyn['t'],dyn['gL'],color='b')
ax2.text(0,23,'Conductances',fontsize=20)
ax2.set_ylim([0,30])
plt.setp(ax2.spines.values(),linewidth=2)
ax2.tick_params(labelsize = 20,width=2,direction='in')
ax2.set_xticklabels([])

# Gating
ax1 = axes[1]
ax1.plot(dyn['t'],dyn['n'],color='k')
ax1.plot(dyn['t'],dyn['m'],color='r')
ax1.plot(dyn['t'],dyn['h'],color='b')
ax1.text(0,0.7,'Gating',fontsize=20)
ax1.set_ylim([0,1])
plt.setp(ax1.spines.values(),linewidth=2)
ax1.tick_params(labelsize = 20,width=2,direction='in')
ax1.set_xticklabels([])

# Voltage
ax0 = axes[0]
ax0.plot(dyn['t'],dyn['V'],color='k',linewidth=3)
ax0.text(0,130,'Membrane voltage (mV)',fontsize=20)
ax0.set_ylim([-20,150])
ax0.text(0,pars['EK'],r'$E_K$',fontsize=20)
ax0.text(0,pars['ENa'],r'$E_{Na}$',fontsize=20)
ax0.text(0,pars['EL'],r'$E_L$',fontsize=20)
plt.setp(ax0.spines.values(),linewidth=2)
ax0.tick_params(labelsize = 20,width=2,direction='in')
ax0.set_xticklabels([])

plt.tight_layout()

return fig
```

SOLUTIONS TO CHALLENGE PROBLEMS

SOLUTION: Completing the HH Setup

The code to insert should replicate the intent of the HH equations, including the nonlinear dependencies of gates as well as the currents as a function of the gating variables and voltages.

```
dyn['gK'] = pars['gKbar']*dyn['n']**4
dyn['gNa'] = pars['gNabar']*dyn['m']**3*dyn['h']
dyn['gL'] = pars['gL']*np.ones(len(dyn['t']))
dyn['IK'] = pars['gKbar'] * dyn['n']**4 * (dyn['V']-pars['EK'])
dyn['INa'] = pars['gNabar'] * dyn['m']**3 * dyn['h'] * (dyn['V']-pars['ENa'])
dyn['IL'] = pars['gL'] * (dyn['V']-pars['EL'])
```

SOLUTION: Completing the HH Equations

The $\dot{V}$ equation as well as the dynamics of gating variables are taken directly from the HH equations and should read as follows:

```
Vdot = (1/pars['C'])*(I-pars['gKbar']*n**4*(V-pars['EK'])\
        -pars['gNabar']*m**3*h*(V-pars['ENa'])-pars['gL']*(V-pars['EL']))
ndot = pars['alphan'](V)*(1-n)-pars['betan'](V)*n
mdot = pars['alpham'](V)*(1-m)-pars['betam'](V)*m
hdot = pars['alphah'](V)*(1-h)-pars['betah'](V)*h
```

It is important to note that in Python the backslash \ represents a formal command; it means that the statement continues on the next line! Hence, although the code above looks strange, it will be accepted by the Python interpreter. However, there can be singularities given the voltage-dependent On/Off rates of gates. The technical appendix includes details on how to interpolate across an apparent singularity in the numerical system of equations. This is implemented in the code available for download.

SOLUTION: Time-Dependent Impulses

Here the code serves double duty. If there is only a single value of `t`, as in the numerical integration, then only a single applied current value will be returned. However, if many values are input, then the code will return the entire function; this is useful for plotting and storing the applied current along with the output.

```
# impulse_t
def impulse_t(t):
    # function I = impulse_t(t)
    # Specifies the applied time-varying current
    # Works if t is a single value or many values

    if isinstance(t,float):
        if t>2 and t<2.5:
            I=2
        elif t>10 and t<10.5:
            I=25
        else:
            I=0
    else:
        I=np.zeros(len(t))
        for i in range(len(t)):
            if t[i]>2 and t[i]<2.5:
                I[i] = 2
```

```
            elif t[i]>10 and t[i]<10.5:
                I[i] = 25
            else:
                I[i]=0
    return I
```

SOLUTION: Customizing Axes in a Single Figure

The following solution combines two features. First, one must deal with the singularity in gate variables to produce a bona fide output without NaN results (NaN = not a number). Second, axes can be stacked in multiple different ways; however, a useful figure code is one that looks the same each time it is run. To make this happen, it is often prudent to specify both the size and layout of the figure window as well as the size and position of each axis. The solution is presented in two parts: first, the calculation elements:

```
require(ggplot2)
require(gridExtra)

# Main data goes here
V=seq(-20,120,0.1)
pars=list()
pars$alphan = function(V) 0.01*(10-V)/(exp(1-V/10)-1)
pars$betan = function(V) 0.125*exp(-V/80)
pars$alpham = function(V) 0.1*(25-V)/(exp(2.5-V/10)-1)
pars$betam = function(V) 4*exp(-V/18)
pars$alphah = function(V) 0.07*exp(-V/20)
pars$betah = function(V) (exp(3-V/10)+1)^-1

# Set the gates
# n
alphan_eval=rep(0,length(V))
tmpi=which(V!=10)
alphan_eval[tmpi]=pars$alphan(V[tmpi])
tmpi=which(V==10)
alphan_eval[tmpi]=0.1

# m
alpham_eval=rep(0,length(V))
tmpi=which(V!=25)
alpham_eval[tmpi]=pars$alpham(V[tmpi])
tmpi=which(V==25)
alpham_eval[tmpi]=1

# h
alphah_eval=pars$alphah(V)
```

Next, we look at axes customization. The layouts will vary by taste. The following code stacks each set of gating variables and compares the α and β values associated with each gate, corresponding to On and Off rates, respectively.

```
fig, axes = plt.subplots(3,1,figsize = (10,15))
# Plot - n gates
axn = axes[2]
axn.plot(V,alphan_eval,color='k',linewidth=3)
axn.plot(V,pars['betan'](V),color=[0.5,0.5,0.5],linewidth=3)
axn.set_ylabel(r'Gates, $\alpha_n$ and $\beta_n$',fontsize=18)
axn.set_xlabel(r'Voltage, $V$',fontsize=18)
axn.legend([r'$\alpha_n$',r'$\beta_n$'],fontsize=16,loc='upper left')
axn.set_yticks(np.arange(0,12,2)*0.1)
plt.setp(axn.spines.values(),linewidth=2)
axn.tick_params(labelsize = 18,width=2,direction='in')

# Plot - m gates
axm = axes[1]
axm.plot(V,alpham_eval,color='k',linewidth=3)
axm.plot(V,pars['betam'](V),color=[0.5,0.5,0.5],linewidth=3)
axm.set_ylabel(r'Gates, $\alpha_m$ and $\beta_m$',fontsize=18)
axm.legend([r'$\alpha_m$',r'$\beta_m$'],fontsize=16,loc='upper left')
axm.set_yticks(np.arange(0,14,2))
axm.set_xticklabels([])
plt.setp(axm.spines.values(),linewidth=2)
axm.tick_params(labelsize = 18,width=2,direction='in')

axh = axes[0]
axh.plot(V,alphah_eval,color='k',linewidth=3)
axh.plot(V,pars['betah'](V),color=[0.5,0.5,0.5],linewidth=3)
axh.set_ylabel(r'Gates, $\alpha_h$ and $\beta_h$',fontsize=18)
axh.legend([r'$\alpha_h$',r'$\beta_h$'],fontsize=16,loc='upper left')
axh.set_yticks(np.arange(0,12,2)/10)
axh.set_xticklabels([])
plt.setp(axh.spines.values(),linewidth=2)
axh.tick_params(labelsize = 18,width=2,direction='in')
```

SOLUTION: Filtering and Excitation

There are different ways to identify the equilibrium. One way is to simulate the model until it relaxes back to steady state and then utilize the equilibrium voltage and gating states as your initial conditions. Another way is to solve for the equilibrium concentrations directly; the key here is to recognize that at equilibrium $V^* = 0$, i.e., there is a zero resting potential. Hence, for the n variable at steady state

$$\alpha_n(V^*)(1-n) = \beta_n(V^*)n, \tag{6.15}$$

which leads to the solution $n^* = \frac{\alpha_n(0)}{\alpha_n(0)+\beta_n(0)}$. Similar logic can be implemented for all the variables as follows:

```
pars['V0']=0
pars['n0']=pars['alphan']*pars['V0']/\
            (pars['alphan']*pars['V0']+pars['betan']*pars['V0'])
pars['m0']=pars['alpham']*pars['V0']/\
            (pars['alpham']*pars['V0']+pars['betam']*pars['V0'])
pars['h0']=pars['alphah']*pars['V0']/\
            (pars['alphah']*pars['V0']+pars['betam']*pars['V0'])
```

SOLUTION: Thresholds for Firing

Here, systematically varying $I(t)$ in increments of 1, identify a switch in behavior from $I = 12$ (left) to $I = 13$ (middle). The firing continues beyond this point, e.g., when $I = 14$ (right).

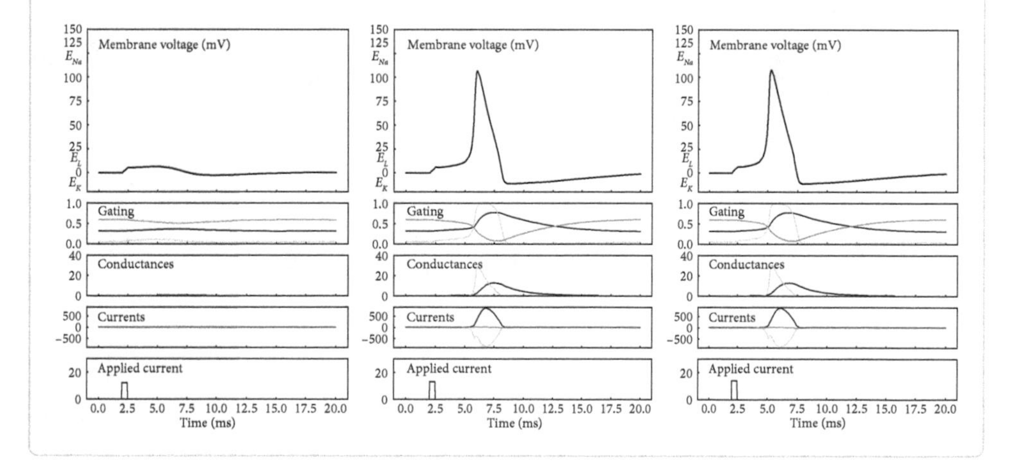

Chapter Seven

Excitations and Signaling, from Cells to Tissue

7.1 EXCITABLE MEDIA: FROM LOCALIZED TO SPATIAL DYNAMICS

Neurons and cardiac cells can respond in an excitable fashion to stimuli. This means that the same cell can have markedly different types of dynamics given the same input, depending on its cellular state—from a dynamical perspective. Thinking about cells as nonlinear dynamical systems reveals that it need not be a structural difference that enables one cell to behave differently than another. Yet understanding the dynamical basis for excitation is complicated.

One barrier is evident from the last chapter. The prior laboratory introduced a computational approach to simulating and examining dynamics emerging in the Hodgkin-Huxley (HH) equations that integrate both fast and slow changes in voltage and voltage-dependent ionic transport channels in neurons. The four-dimensional, nonlinear system of ODEs is a challenge to analyze. Yet, its behavior is readily apparent. Dynamics include filtering of small impulses, excitatory responses to large impulses beyond a threshold, and refractory periods during excitation (where the system seems to "ignore" a second impulse). There are other dynamics that can also be investigated, e.g., regular beating dynamics that arise when given a constant stimulus. Is it possible to recapitulate such features in a simpler model? And, if so, might it also be possible to understand more about the HH dynamics by investigating a simpler system?

This lab tries to answer these questions affirmatively through the study of what is, in essence, a model of a model. The goal is to develop an understanding of oscillatory dynamics observed in excitable media, such as spiking neurons and cardiac cells. We focus on a simplified model of voltage spiking: the FitzHugh-Nagumo model, or FN model for short (FitzHugh 1961; Nagumo et al. 1962). The FN model is a model of the HH model. It is intentionally simplified yet exhibits many of the same features. Using the FN model, it is possible to explore the voltage dynamics of single cells and recapitulate some key phenomena, including filtering and excitatory dynamics. However, we also strive to do more. Individual neurons and cardiac cells are connected to each other. Hence, dynamics at one cell can spread, whether via axons or by self-excitatory waves through tissue. As a first step toward understanding how to move from one cell to many, this lab also introduces a model of cardiac dynamics at the tissue scale using a new approach: that of partial differential equations (PDEs). The lab focuses on a simplified case of one-dimensional excitations as a means to understand how excitations in one location can propagate and modify the behavior of many cells and even tissue.

Part 1 of the lab begins with the FN model of firing. The model is depicted in the schematic below. It includes a (potentially) fast voltage response and a slow gating response. In part 2, individual cells are connected such that voltage can diffuse between cells. This diffusion can enable a new excitation at a nearby cell. Paradoxically, the combination of excitation and diffusion can lead to a new phenomenon: that of a traveling wave that could represent the spread of a cardiac impulse through tissue or the transmission of an action potential down an axon in a brain network. It's time to begin.

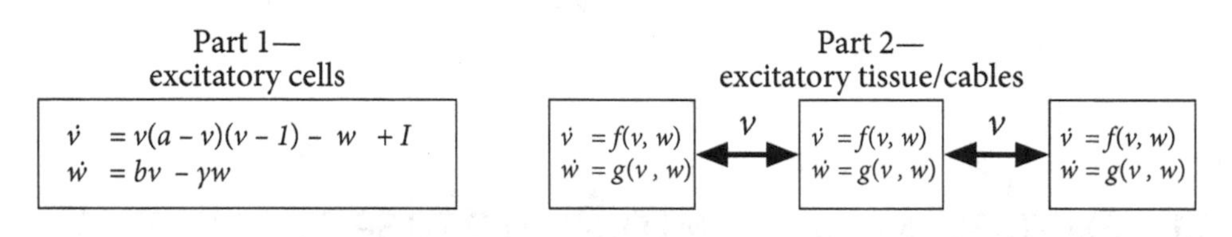

7.2 FITZHUGH-NAGUMO: THE ODE MODEL

Information is transmitted between neurons in the form of spikes—action potentials. The FitzHugh-Nagumo (FN) model is a simple model for the firing activity of a single neuron or cardiac cell. There are different variants, and we use the following:

$$\dot{v} = v(a-v)(v-1) - w + I(t)$$

$$\dot{w} = bv - \gamma w$$

where v is the membrane voltage and w is the response. The parameter $I(t)$ represents a time-dependent current that may be set to 0. The other parameters are biophysically motivated as they are combinations of parameters from the Hodgkin-Huxley model from which the FN model is derived.

The dynamics of the FN model can be understood, in part, by analyzing the shape of the nullclines. The $\dot{v}=0$ nullcline is a cubic equation and the $\dot{w}=0$ nullcline is a linear equation in phase space.

$$\dot{v} = 0 \rightarrow w = I + v(a-v)(v-1)$$

$$\dot{w} = 0 \rightarrow w = \frac{b}{\gamma} v$$

These nullclines form the basis for the first challenge problem.

CHALLENGE PROBLEM: Nullclines of the FN Model

Plot the nullclines for $a=0.1$, $b=0.01$ and $\gamma=0.02$, $I=0$, as seen in the image below. The two nullclines cross at $v=0$ and $w=0$, which is an equilibrium of the dynamics, but is it stable? Why or why not? Part of the core objective of this lab is to answer this question and to understand the difference between dynamics that follow small versus large perturbations.

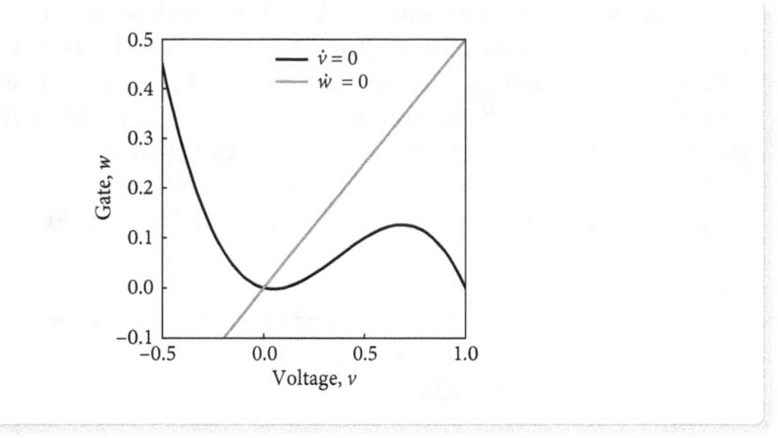

7.2.1 Filtering small perturbations

As is evident from the prior section, this version of the FN model has an equilibrium at $(0, 0)$. Is this equilibrium stable? Until this point, we have defined stability in terms of *small changes* in densities, i.e., local stability. Given this 2D system, it is possible to assess the stability analytically. However, let's begin by analyzing the stability of the equilibrium by perturbing the system with a small change: $v = 0.1$ and $w = 0$. To do so, write up your own code, or use the following code snippet to simulate the model with Python's `integrate.odeint` until $t = 1000$:

```
def model_fn(x,t,pars):
    v = x[0]
    w = x[1]

    dxdt = [v*(pars['a']-v)*(v-1)-w+pars['I'], pars['b']*v-pars['gamma']*w]
    return dxdt
```

This code uses the convention that parameters are passed with the structure `pars`. The dynamics are as follows:

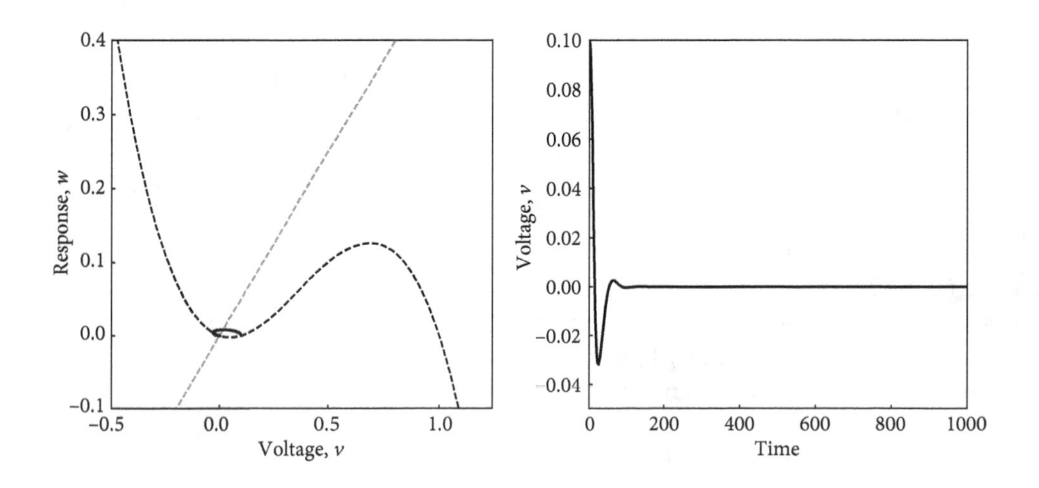

As is apparent, the equilibrium is stable, at least with respect to certain small perturbations. The code sets the relative tolerance, `rtol`= 10^{-6}, because the dynamics can be "stiff." A stiff dynamical system is one characterized by dynamics that have regions over which certain variables change very rapidly given typical time steps. And, in other cases, variables may change very slowly. The rapid changes often require solvers to reduce step sizes to very low levels in order to maintain accuracy.

The code snippet for assessing perturbations of the FN model is given below.

```
# Simulation of the FN model
# The code sets integration options to manipulate the relative
# tolerance of the numerical integration
# Both subplots show dynamics (left-phase, right-time)
import numpy as np
import matplotlib.pyplot as plt
from scipy import integrate

pars = {}
pars['a'] = 0.1
pars['b'] = 0.01
pars['gamma'] = 0.02
pars['I'] = 0.0

y0=[0.1, 0]
t = np.linspace(0,1000,1000)
y = integrate.odeint(model_fn,y0,t,args=(pars,),rtol=10**-6)

# Plotting components
fig, axes = plt.subplots(1,2, figsize = (15,7))

ax0=axes[0]
v = np.arange(-5, 5.01, 0.01)
wnull_v = v*(pars['a']-v)*(v-1)+pars['I']
wnull_w = pars['b']*v/pars['gamma']

ax0.plot(v,wnull_v,color='k',linestyle='--',linewidth=2)
ax0.plot(v,wnull_w,color=[0.5,0.5,0.5],linestyle='--',linewidth=2)
ax0.plot(y[:,0],y[:,1],color='k',linewidth=3)

ax0.set_ylim([-.1, 0.4])
ax0.set_xlim([-.5, 1.25])
ax0.set_xlabel(r'Voltage, $v$',fontsize=20)
ax0.set_ylabel(r'Response, $w$',fontsize=20)
ax0.set_xticks(np.linspace(-0.5,1,4))
plt.setp(ax0.spines.values(),linewidth=2)
```

```
ax0.tick_params(labelsize=14,direction='in',width=2)

ax1=axes[1]
ax1.plot(t,y[:,0],linewidth=3,color='k')
ax1.set_xlabel('Time',fontsize=20)
ax1.set_ylabel(r'Voltage, $v$',fontsize=20)
ax1.set_xlim([0,1000])
ax1.set_ylim([-0.05,0.1])
plt.setp(ax1.spines.values(),linewidth=2)
ax1.tick_params(labelsize=14,direction='in',width=2)

plt.tight_layout()
```

7.2.2 Single action potential

What happens in the FN model if the cell is exposed to a larger perturbation? To answer this question, change the simulation such that the initial condition is $v = 0.2$ and $w = 0$. The resulting dynamics should look like the following:

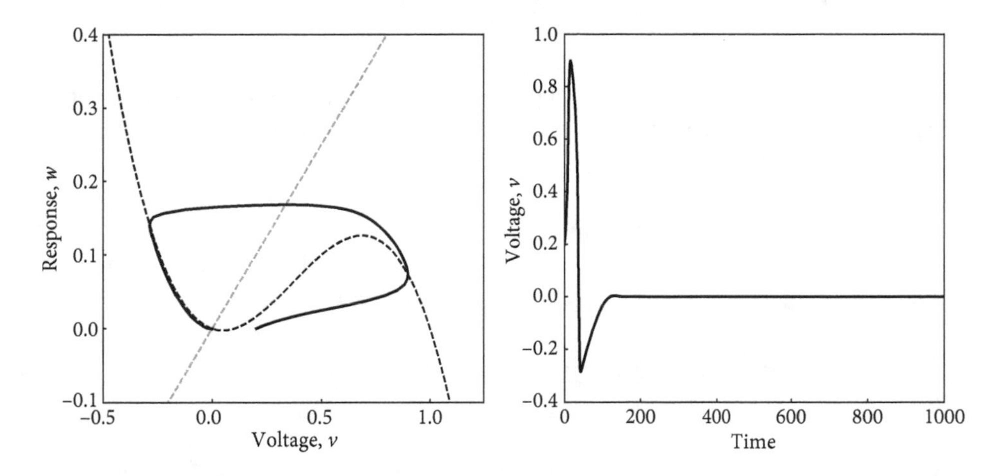

The equilibrium is still stable—in a global sense; however, the dynamics do not immediately relax to the equilibrium. For these parameters, a sufficiently large perturbation leads to a single action potential.

CHALLENGE PROBLEM: Critical Transition to Excitability

Vary v until you identify a critical perturbation size at which point the system excites, rather than relaxes back to equilibrium. In the interest of time during the lab, consider varying the initial value in increments of $\Delta v = 0.01$, or if you have the bandwidth, try $\Delta v = 0.001$.

7.2.3 Sustained oscillations

The FN model can also exhibit sustained oscillations, aka beating. An example of beating dynamics can be seen by increasing the value of the parameter I, analogous to the applied current in the HH model. To begin, increase I to 0.15. Biologically, this corresponds to an injection of current. Again, plot the nullclines.

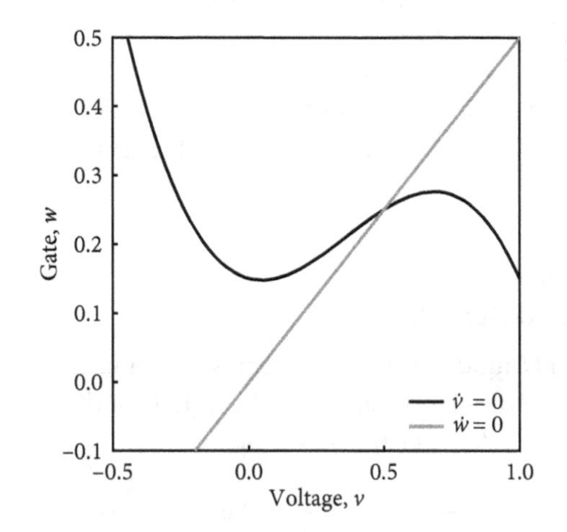

The intersection of the nullclines is no longer at $(0,0)$. Instead, the new equilibrium appears to be located near $(0.5, 0.25)$ in the phase plane. This can be solved formally by identifying the intersection of the nullclines.

```
from scipy import optimize

wstarfun = lambda v: pars['b']/pars['gamma']*v
vstarfun = lambda v: pars['I']+v*(pars['a']-v)*(v-1)-wstarfun(v)
vstar = optimize.fsolve(vstarfun,0)
wstar = wstarfun(vstar)
```

As you can see, the value of the equilibrium is in fact $(v^* = 0.5, w^* = 0.25)$. Here w^* was identified using $\dot{w} = 0$ and then substituting the results into $\dot{v} = 0$ to obtain the equilibrium. It is now possible to determine the local stability of the equilibrium by computing the eigenvalues of the Jacobian. The Jacobian matrix is

$$\begin{pmatrix} -3v^2 + 2(1+a)v - a & -1 \\ b & -\gamma \end{pmatrix}.$$

Below is a handy function that outputs the equilibrium and the eigenvalues of the Jacobian for the FN model given parameters in `pars`.

```
def FNequil(pars):
  # function [xstar,eigvals] = FNequil(pars)
  # Returns both the equilibrium and the eigenvalues of the
  # FN model for the parameters in "pars"
```

```
    wstarfun = lambda v: pars['b']/pars['gamma']*v
    vstarfun = lambda v: pars['I']+v*(pars['a']-v)*(v-1)-wstarfun(v)
    # Get equilibrium values. Note: fsolve returns an array, so we
    # must unpack that with the [0] at the end
    vstar = optimize.fsolve(vstarfun,0)[0]
    wstar = wstarfun(vstar)

    xstar = np.array([vstar, wstar])

    eigmatfun = lambda v,w: [[-3*v**2+2*(1+pars['a'])*v-pars['a'], -1],
                              [pars['b'],-pars['gamma']]]
    eigvals, eigvecs = np.linalg.eig(eigmatfun(vstar,wstar))
    return [xstar, eigvals]
```

With this script in hand, it is possible to predict what will happen given small perturbations away from the equilibrium, as in the next challenge problem.

CHALLENGE PROBLEM: Stability of the FN Model with a Baseline Current

What is the equilibrium value and what are the eigenvalues of the FN model for the set of default parameters given $I = 0.15$? Compare and contrast the dynamics to the solutions when $I = 0$, e.g., by plotting the dynamics in the case when $I = 0.15$.

7.3 FITZHUGH-NAGUMO: ONE-DIMENSIONAL PDEs

7.3.1 Overview

The previous model was appropriate for a single excitable cell. However, brain and cardiac tissue is composed of spatially distributed excitable cells that can excite each other. By exciting a neighboring cell, voltage appears to diffuse from one cell to the other. This can be modeled by including diffusion in the voltage equation:

$$\frac{\partial v}{\partial t} = v(a-v)(v-1) - w + I + D\nabla^2 v \tag{7.1}$$

$$\frac{\partial w}{\partial t} = bv - \gamma w \tag{7.2}$$

In this case, the Laplacian operator, ∇^2, is equivalent to $\frac{\partial^2}{\partial x^2}$ in one dimension, $\frac{\partial^2}{\partial x^2} + \frac{\partial^2}{\partial y^2}$ in two dimensions, and so forth. We solve this equation numerically for a particular geometry of cells: a one-dimensional cable. This is the first step toward a full-fledged spatial simulation of excitable dynamics in living cells. To get there, it is critical to first develop the theory and then introduce the code. This code is essential to the implementation of the laboratory and is available online.

7.3.2 Approximation method

Numerical solutions to PDEs typically involve discretizing space and solving for the time evolution using finite difference methods. The specifics of the finite difference method depend on the geometries used in the problem and the desired accuracy of the results. In our example of a one-dimensional "cable" of cells, we use the simplest finite difference method—the explicit Euler method. This means we must apply the explicit Euler method to approximate $\frac{\partial v}{\partial t}$, $\frac{\partial w}{\partial t}$, and $\frac{\partial^2 v}{\partial x^2}$. A visualization of the Euler method applied to solving the voltage cable of length L is shown below.

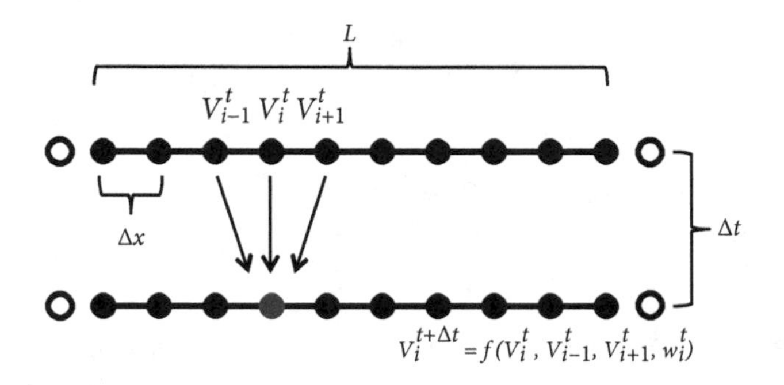

Here the cable has been discretized (filled circles) with a distance Δx between points. These points allow us to calculate appropriate estimates to the spatial derivatives. Additionally, we ascribe values to points outside of the cable (unfilled circles). These "ghost cells" help when estimating $\frac{\partial^2 v}{\partial x^2}$.

The time evolution proceeds in discrete steps, Δt, allowing for a finite difference approximation of the time derivative for each specific point of the cable. Mathematically, a point is updated over time:

$$\frac{dv_i}{dt} \approx \frac{v_i^{t+\Delta t} - v_i^t}{\Delta t}$$

with a similar equation for the w variable. We see the voltage at one point, i, results from contributions due to the nondiffusive and diffusive terms in the model.

$$\frac{\partial v}{\partial t} = v(a-v)(v-1) - w + I + D\nabla^2 v = f(v,w) \implies \tag{7.3}$$

$$v_i^{t+\Delta t} = v_i^t + \Delta t f(v,w) \tag{7.4}$$

How accurate is this approximation? To address this question, we show that this (and the other finite difference approximations) are derived from a Taylor expansion:

$$v(t+\Delta t) = v(t) + \frac{dv}{dt}\Delta t + \frac{1}{2}\frac{d^2 v}{dt^2}\Delta t^2 + \cdots \tag{7.5}$$

For this approximation, we ignore terms of order Δt^2 and higher. Meanwhile, the error of this approximation for $V_i^{t+\Delta t}$ is of order $O(\Delta t)$. That is, the error grows linearly with increases in the time step. The error is linear because we ignore terms of Δt^2, but we divide by Δt for the update rule to $v_i^{t+\Delta t}$.

The contribution from the nondiffusive terms in $f(v, w)$ are calculated from the values at each lattice point, v_i^t and w_i^t. The diffusive terms include contributions from the focal cell, i, and the neighboring cells. The diffusive term is also estimated by applying a finite difference approximation of the Laplacian, ∇^2. The Taylor expansion about x up to second order is

$$v_{i\pm 1} = v(x \pm \Delta x) \approx v(x) \pm \frac{\mathrm{d}v}{\mathrm{d}x}\bigg|_x \Delta x + \frac{1}{2}\frac{\mathrm{d}^2 v}{\mathrm{d}x^2}\bigg|_x \Delta x^2.$$

If we add $v_{i+1} + v_{i-1}$, all odd terms cancel, leading to

$$v_{i+1} + v_{i-1} \approx 2V(x) + \frac{\mathrm{d}^2 v}{\mathrm{d}x^2}\bigg|_x \Delta x^2 \Longrightarrow$$

$$\frac{\mathrm{d}^2 v}{\mathrm{d}x^2} \approx \frac{v_{i-1} - 2v_i + v_{i+1}}{\Delta x^2}$$

Note the error is $O(\Delta x^2)$ because the next nonzero term is of order Δx^4; dividing by Δx^2 in the approximation would eliminate such higher-order terms.

Inputting these two finite difference approximations into the system of equations yields

$$\frac{v_i^{t+\Delta t} - v_i^t}{\Delta t} = v_i^t(a - v_i^t)(v_i^t - 1) - w_i^t + I + D\frac{v_{i-1} - 2v_i + v_{i+1}}{\Delta x^2}$$

$$\frac{w_i^{t+\Delta t} - w_i^t}{\Delta t} = bv_i^t - \gamma w_i^t$$

from which we can easily solve for the updated $v_i^{t+\Delta t}$ and $w_i^{t+\Delta t}$.

The relationship above does not apply to the end points of the cable. The end points evolve depending on the boundary conditions. Boundary conditions dictate mathematical conditions that must be satisfied at the end point and are prescribed by the biological circumstance. For our example of a finite cable of excitable cells, the voltage has no way to leave the system (you can think of the cable being secured at the lab bench by an insulator like glass). In this case, no-flux boundary conditions $\frac{\mathrm{d}v}{\mathrm{d}x} = 0$ are used at the edges of the tissue. This boundary condition is approximated by a finite difference implementation. In order to obtain an approximation, we return to the Taylor expansion:

$$v_{i\pm 1} = v(x \pm \Delta x) \approx v(x) \pm \frac{\mathrm{d}V}{\mathrm{d}x}\bigg|_x \Delta x + \frac{1}{2}\frac{\mathrm{d}^2 v}{\mathrm{d}x^2}\bigg|_x \Delta x^2$$

$$v_{i+1} - v_{i-1} \approx 2\frac{\mathrm{d}v}{\mathrm{d}x}\bigg|_x \Delta x \Longrightarrow$$

$$\frac{\mathrm{d}v}{\mathrm{d}x}\bigg|_x \approx \frac{v_{i+1} - v_{i-1}}{2\Delta x}$$

where the last line is $O(\Delta x^3)$. The even terms are eliminated in the second line, leaving terms proportional to Δx^3 as the lowest order of ignored terms. Again, we divide by Δx in the approximation. From the approximation, the condition that must be satisfied at the leftmost boundary where $x = 0$ is

$$\left.\frac{dv}{dx}\right|_x = 0 \Longrightarrow \frac{v_2 - v_0}{2\Delta x} = 0.$$

Of course, the value at $x = 0$ is not defined. Thus, in the discretization of space, we track the values at $x = 0$ and $x = N + 1$ such that

$$v_0 = v_2$$

$$v_{N+1} = v_{N-1}$$

These are the ghost cells in the schematic. Their role is to enforce the no-flux boundary conditions and maintain the numerical accuracy of the approximations.

With this mathematical background, we can implement the code. First, set the parameters to the same values we used in the ODE FitzHugh-Nagumo model.

```
# Length of cable
Lx = 100

# Number of points for approximation
Nx = 100
dx = Lx/Nx

# Total cells including ghost cells
totN = Nx + 2

#Total time of dynamics
T=100

# Parameters from ODE
pars={}
pars['a'] = 0.1
pars['b'] = 0.01
pars['gamma'] = 0.02
pars['I'] = 0
pars['D'] = 1 # Diffusion constant
pars['dx'] = 1

dt = 0.05
totruns = int(np.round(T/dt))
# Preallocate
V = np.zeros(totN)
w = np.zeros(totN)
```

```
data = {}
data['allv'] = np.zeros((Nx,totruns+1))
data['allw'] = np.zeros((Nx,totruns+1))
data['tvec'] = np.arange(0,T+dt,dt)
```

Note that here we set $\Delta t = 0.05$. In order for diffusion to be stably approximated, the time increment should satisfy the following condition: $\Delta t < \frac{\Delta x^2}{2D}$. For the values above, that corresponds to $\Delta t < 0.5$, and we choose a value 1/10-th of that. Here numerical stability means errors do not grow over time. Stability is a necessary condition for good numerical approximations. We choose to fix Δt to a value smaller than the limit required for stability in order to reduce the error (recall the error is linear in Δt). The `data` structure will contain the dynamics along the cable (no ghost cells) across the entire simulation, while the v and w vectors will track the updated state (including ghost cells) over time. Based on the schematic, the main skeleton of the code loops over time while updating the state at each spatial location at each time, as seen in the solution to the next challenge problem.

CHALLENGE PROBLEM: A 1D FN Model

Modify the following code to simulate the full 1D FN model of an excitable tissue.

```
# Partially operational FN model - to be filled in
t = 0
cnt = 1
while t<T:
    # Calculate terms in PDE

    # Update with time

    # Fix boundary conditions for no-flux

    # Update time
    t = t+dt
    cnt = cnt+1
```

Let's break this down . . . piece by piece. First, the terms calculated are the "reactions," i.e., the FN model dynamics, and diffusion:

```
rxnV = V*(pars['a']-V)*(V-1)-w+pars['I']
diffV = pars['D']*(V[:-2]-2*V[1:-1]+V[2:])/dx**2
rxnw = pars['b']*V-pars['gamma']*w
```

where vector notation simplifies the calculation, particularly for the diffusion term. These terms combine to update the v equation.

```
V[1:-1] = dt*(rxnV[1:-1]+diffV) + V[1:-1]
```

Note that this only updates the non-ghost cells. Last, we must update the value of the ghost cells. From before we had

$$v_0 = v_2$$

$$v_{N+1} = v_{N-1}$$

These boundary conditions must be implemented to ensure that the ghost cells of the V variable are updated every time you increment. *Be careful of your indices here if you are developing your own simulation!!!* Note that you don't have to update the w ghost cells because there is no diffusion in that equation. However, the code above does so for generality.

7.3.3 Spatial dynamics of the 1D FN model

We first consider perturbations in voltage in the 1D FN model. Previously, a perturbation of $\Delta v = 0.2$ caused a single action potential. What happens in space? We can test this out by varying the initial condition of the code to include a perturbation to the eleventh cell in the cable.

```
V[10] = 0.2
```

Run the function and watch a movie of the dynamics using the following code. We want the movie to play out in its own window, so make sure to give the command "%matplotib qt" in the console before running the code.

```
fig, axes = plt.subplots(1,2, figsize = (15,7))
ax0=axes[0]
ax1=axes[1]

for j in range(totruns+1):
    ax0.clear()
    currv = data['allv'][:,j]
    currw = data['allw'][:,j]
    ax0.plot(np.arange(Nx),currv,linewidth=3,color='k')
    ax0.set_xlabel('Cell in Cable',fontsize=20)
    ax0.set_ylabel(r'Voltage, $v$',fontsize=20)
    ax0.set_ylim(np.min(np.min(data['allv'])), np.max(np.max(data['allv'])))
    ax0.set_xlim(0,Nx)
    plt.setp(ax0.spines.values(),linewidth=2)
    ax0.tick_params(labelsize=14,direction='in',width=2)

    ax1.clear()
    ax1.plot(np.arange(Nx),currw,linewidth=3,color='k')
    ax1.set_xlabel('Cell in Cable',fontsize=20)
    ax1.set_ylabel(r'Response, $w$',fontsize=20)
    ax1.set_ylim(np.min(np.min(data['allw'])), np.max(np.max(data['allw'])))
```

```
ax1.set_xlim(0,Nx)
plt.setp(ax1.spines.values(),linewidth=2)
ax1.tick_params(labelsize=14,direction='in',width=2)

plt.tight_layout()
plt.pause(0.01)
```

The dynamics below show that, for the spatial model, the perturbation does not lead to excitation. That's because the voltage diffuses before activating the action potential for any one cell in the cable.

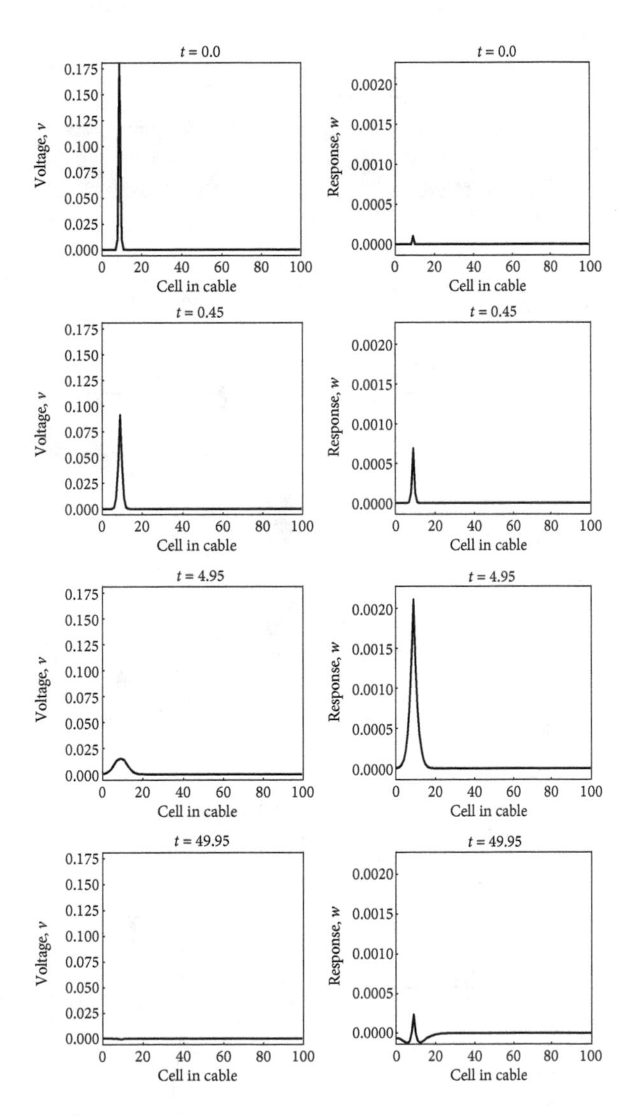

Now let's try a larger perturbation:

```
V[10]=2
```

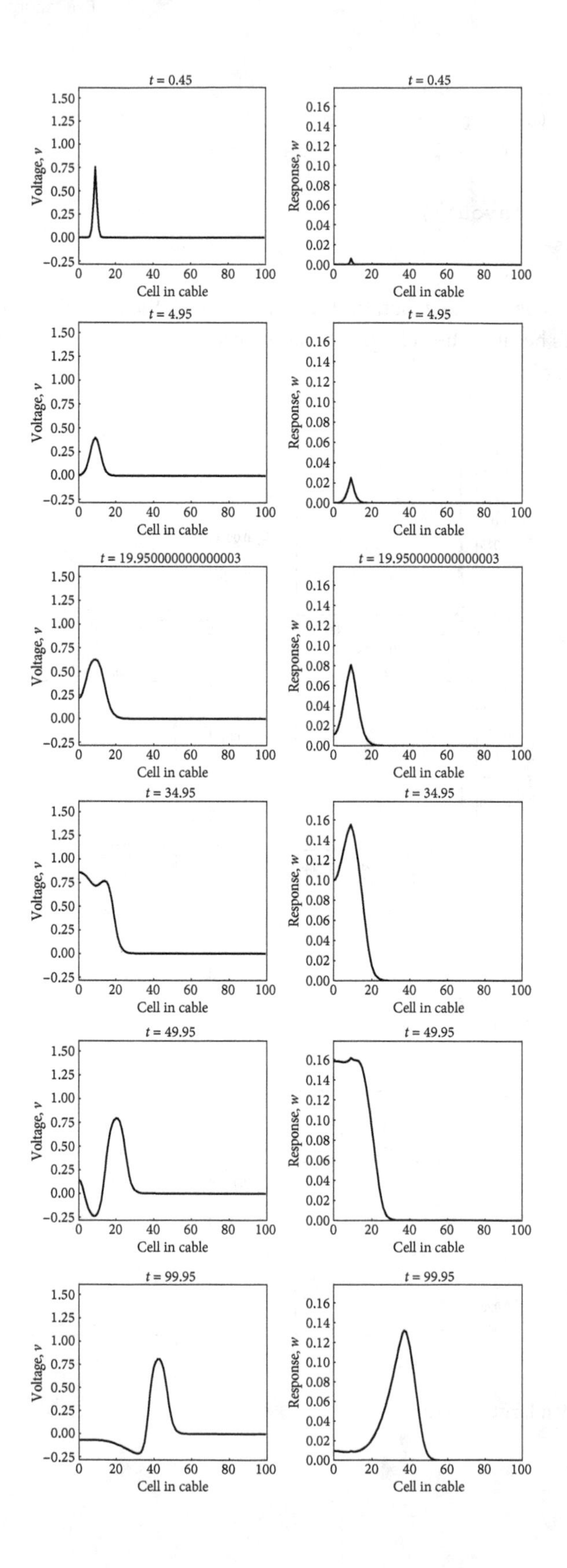

t = 0.45
Voltage, v
Cell in cable
t = 0.45
Response, w
Cell in cable
t = 4.95
Voltage, v
Cell in cable
t = 4.95
Response, w
Cell in cable
t = 19.950000000000003
Voltage, v
Cell in cable
t = 19.950000000000003
Response, w
Cell in cable
t = 34.95
Voltage, v
Cell in cable
t = 34.95
Response, w
Cell in cable
t = 49.95
Voltage, v
Cell in cable
t = 49.95
Response, w
Cell in cable
t = 99.95
Voltage, v
Cell in cable
t = 99.95
Response, w
Cell in cable

The voltage initially drops, but it remains high enough to stimulate an action potential. This action potential then stimulates the neighboring cells via diffusion. This results in a (slow-moving) traveling wave.

This is the first step toward a spatial simulation of excitable media. Not bad for one day in the laboratory. The homework builds upon these routines and helps you get to the next step, e.g., exploring wave speeds, simulating axonal dynamics, and perhaps even studying excitable waves in cardiac tissue.

SOLUTIONS TO CHALLENGE PROBLEMS

SOLUTION: Nullclines of the FN model

The solution to this challenge problem involves plotting both the $\dot{v}$ and $\dot{w}$ dependencies on the same axis. As noted, there is a single intersection for this parameter combination at $(0, 0)$. The stability of the fixed point takes more time to analyze.

```
import numpy as np
import matplotlib.pyplot as plt

# Parameters
pars = {}
pars['a'] = 0.1
pars['b'] = 0.01
pars['gamma'] = 0.02
pars['I'] = 0.0

# Nullclines
v = np.arange(-5, 5.01, 0.01) # Range for v
wnull_v = v*(pars['a']-v)*(v-1)+pars['I']
wnull_w = pars['b']*v/pars['gamma']

# Visualization
fig = plt.figure(figsize=(5,5))
ax = fig.gca()

plt.plot(v,wnull_v,color='k',linewidth=3)
plt.plot(v,wnull_w,color=[0.5,0.5,0.5],linewidth=3)
plt.ylim([-.1,.5])
plt.xlim([-.5, 1])
plt.ylabel(r'Gate, $w$',fontsize=20)
plt.xlabel(r'Voltage, $v$',fontsize=20)
plt.xticks(np.linspace(-0.5,1,4))
plt.legend([r'$\dot{v}=0$',r'$\dot{w}=0$'],loc='upper center',
           frameon=False,fontsize=14)
plt.setp(ax.spines.values(),linewidth=2)
ax.tick_params(labelsize=14,direction='in',width=2)
```

SOLUTION: Critical Transition to Excitability

The following code varies the initial perturbation, simulates the FN model, calculates the maximum values of the excursion, and then compares this maximum to the initial perturbation. As one can see in the code and results below, there is a rapid transition somewhere between $v_0 = 0.162$ and $v_0 = 0.163$, where the system has a small increase and then decline compared to a very large increase corresponding to an excitation in the system. This can be seen in two ways: first, on the left is the maximum voltage, which seems to scale linearly with v_0 until it rapidly transitions to over 0.8; second, on the right the time dynamics of the voltage reveal the divergent trajectories given a 0.001 change in v_0. One additional point to consider: The initial voltage of $v_0 = 0.163$ starts the dynamics on the right side of the v nullcline in phase space. Hence, is it possible to find the critical initial voltage for excitability by identifying the intersection of the v axis with the nullcline?

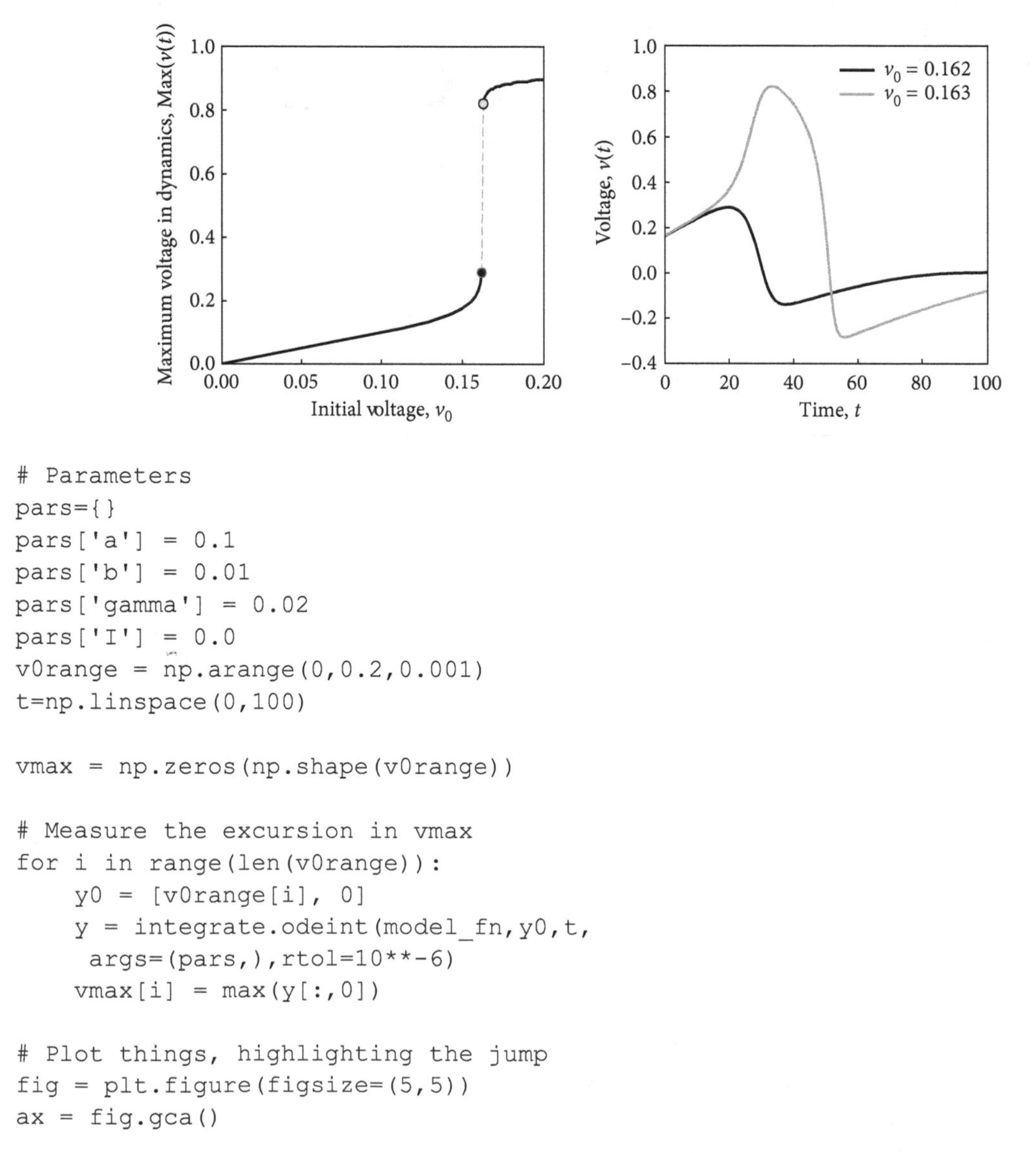


```
# Parameters
pars={}
pars['a'] = 0.1
pars['b'] = 0.01
pars['gamma'] = 0.02
pars['I'] = 0.0
v0range = np.arange(0,0.2,0.001)
t=np.linspace(0,100)

vmax = np.zeros(np.shape(v0range))

# Measure the excursion in vmax
for i in range(len(v0range)):
    y0 = [v0range[i], 0]
    y = integrate.odeint(model_fn,y0,t,
     args=(pars,),rtol=10**-6)
    vmax[i] = max(y[:,0])

# Plot things, highlighting the jump
fig = plt.figure(figsize=(5,5))
ax = fig.gca()
```

```
plt.plot(v0range[:163],vmax[:163],color='k',
 linewidth=3)
plt.plot(v0range[163:],vmax[163:],color='k',
 linewidth=3)
plt.plot(v0range[162:164],vmax[162:164],
        color=[0.5,0.5,0.5],linewidth=2,linestyle='--')
plt.plot(v0range[162],vmax[162],marker='o',markerfacecolor='k',markersize=8)
plt.plot(v0range[163],vmax[163],marker='o',markerfacecolor=[0.75,0.75,0.75],
        markeredgecolor='k',markersize=8)
```

SOLUTION: Stability of the FN Model with a Baseline Current

The eigenvalues (0.206) and (0.024) are both positive, meaning the equilibrium is locally unstable. This contrasts with the case when $I = 0$ where the real values of both complex eigenvalues are negative. The nullclines feature only one intersection and therefore only one equilibrium. The figure below shows the dynamics given initial conditions of $(0.5, 0.26)$.

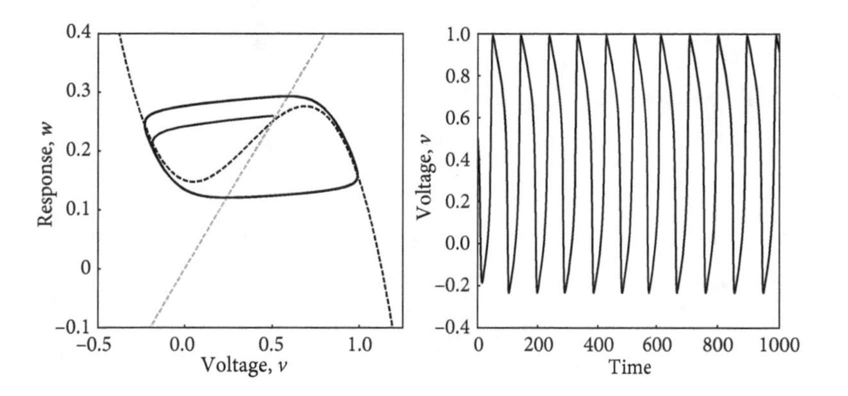

The dynamics feature sustained action potentials. This forms a periodic orbit. Take the last point of the dynamics and perturb slightly from it and observe the dynamics. The dynamics return to the periodic orbit. This periodic orbit is referred to as a *limit cycle*. Like equilibrium points, limit cycles can be stable (as observed here) or unstable. A stable limit cycle attracts points in the neighborhood to it. In contrast, the distance of points in phase space near an unstable limit cycle begins to increase, such that the dynamics do not return to the limit cycle.

SOLUTION: A 1D FN Model

The code below provides the solution, i.e., a fully operational Death Star. Oh wait, that was in a certain science fiction movie. This is a fully operational 1D FN model—much nicer to have in the galactic neighborhood.

```
t = 0
cnt = 1
while t<T:
    # Calculate terms in PDE
    # Reaction term of V in FN
    rxnV = V*(pars['a']-V)*(V-1)-w+pars['I']
    # Voltage diffusion
    diffV = pars['D']*(V[:-2]-2*V[1:-1]+V[2:])/dx**2
    # reaction term of w in FN
    rxnw = pars['b']*V-pars['gamma']*w

    # Evolve over time
    V[1:-1] = dt*(rxnV[1:-1]+diffV) + V[1:-1]
    w[1:-1] = dt*(rxnw[1:-1]) + w[1:-1]
    data['allv'][:,cnt] = V[1:-1]
    data['allw'][:,cnt] = w[1:-1]

    # Fix boundary conditions for no-flux
    V[0] = V[2]
    V[-1] = V[-3]
    # Not necessary for w, but we'll do it anyway for consistency
    w[0] = w[2]
    w[-1] = w[-3]

    # Update time
    t = t+dt
    cnt = cnt+1
```

Chapter Eight

Organismal Locomotion through Water, Air, and Earth

8.1 INTRODUCTION

This laboratory culminates Part II and our study of organismal behavior and physiology. It comes full circle, in a sense, from the Chapter 5 laboratory that covered *E. coli* dynamics from the perspective of chemotaxis. There the objective was to represent the kinase cascade that transformed external signals into changes in the run and tumble motion of the microbe. Hence, the lab focused largely on how microbial cells modulate their behavior in light of environmental gradients. Yet the behavior was taken for granted. Somehow *E. coli* moves. Likewise, many other microbes, cells, and organisms at scales from microns to meters in length manage to explore their environments with gaits, behaviors, and movements that often share surprisingly common principles. This chapter takes inspiration from the humble *E. coli* and focuses on one particular limit: when damping (aka resistive; aka viscous) forces are so large that they overwhelm the inertial forces. In doing so, it is possible to understand how organisms move, by generating internal forces that change their own body space configuration (and center of mass), thereby interacting with the environment.

To accomplish this objective, the lab centers on Purcell's (1977) three-link swimmer (Figure 8.1). The three-link swimmer has two internal degrees of freedom, α_1 and α_2, denoting the angles of the two side links relative to the center link. The angles of these links are controlled by the swimmer, and we assume that the swimmer always has sufficient force to (precisely) control changes in its internal configuration, i.e., $(\alpha_1(t), \alpha_2(t))$. If these internal dynamics repeat, i.e., $\alpha_1(t) = \alpha_1(t+T)$ and $\alpha_2(t) = \alpha_2(t+T)$, we can call such periodic orbits a "gait" in configuration space. But orbits in configuration spaces don't tell the complete story. The objective of this laboratory is to connect shape deformations to motion in the world frame—what is denoted as (x, y, θ) for the three-link swimmer in Figure 8.1. By focusing on the humble three-link swimmer, this laboratory is a gateway to a larger question: understanding how an organism that moves its flagella, undulates its body, or flaps its wings manages to do more than just flail helplessly but instead to swim, dig, or soar into flight.

This field is mathematically rich and challenging. The conventional approach is to consider the zero-inertia limit in which an organism cannot move unless it continues to generate internal body forces, i.e., the no-coasting limit. In that sense, one must solve for

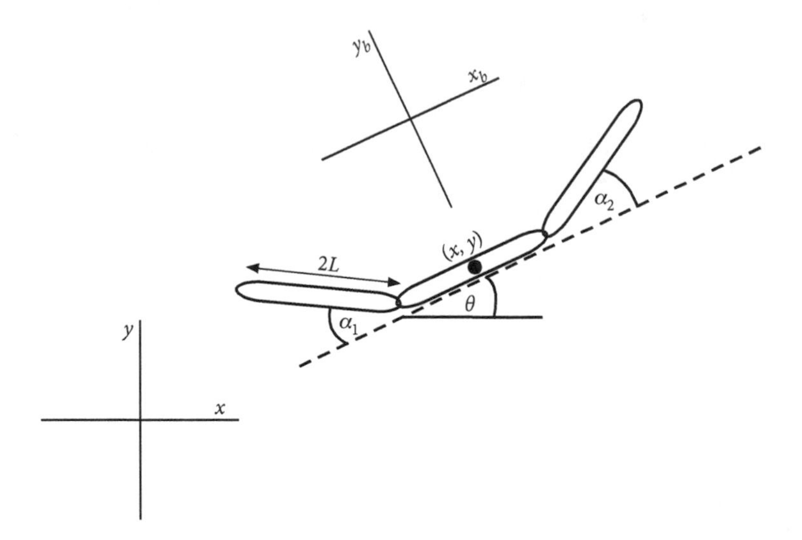

Figure 8.1: Three-link swimmer with limbs that can move given a contact point to the body via two angles, α_1 and α_2. The angle of the swimmer in the world frame is θ. The centerpoint position of the swimmer in the world frame is (x, y). In the body frame, any point on the swimmer can be designated by the coordinates (x_b, y_b), such that in the world frame all points can be written as $(x + x_b, y + y_b)$.

the steady state velocity consistent with the absence of inertial forces given a particular gait. Doing so opens the door to notions of geometric phase and other profound insights that transcend the particular system (i.e., the ideas hold for microbes swimming in water as they do for metazoans "swimming" in sand). However, this path requires an understanding of Navier-Stokes equations (Tritton 1988) as well as a familiarity with notions of gauge theory (Shapere and Wilczek 1989). Instead, we take a dynamical systems perspective and develop a computational model that incorporates basic Newtonian physics (think $F = ma$) with internal gait programs (think $\alpha_i(t)$) as a means to connect internal force generation with movement in the world frame. As described in the textbook, this lab bridges the paradigms of Newton and Borelli, and hopefully brings simple physical ideas of dynamical motion to life.

8.2 THE INTERNAL ORIGINS OF MOVEMENT

Consider a three-link swimmer whose position and orientation in the world frame are described by (x, y, θ), where x and y denote the position of the centerpoint of the body and θ denotes the angle of the center link relative to the x axis. Hence, there are many configurations of the swimmer irrespective of where it is located in space or how it is rotated. The internal configuration space represents the menagerie of shapes that the three-link swimmer can adopt. For example, when both $\alpha_1 = -\pi/2$ and $\alpha_2 = -\pi/2$, the configuration looks likes a paper clip, while when $\alpha_1 = -\pi/2.5$ and $\alpha_2 = \pi/2$, the configuration looks like someone in a reclining chair. Indeed, viewing these configurations is central to the concept of gaits, which represent orbits in configuration space and therefore shifts in internal body shape—i.e., the positioning of links relative to one another.

Let's see how. Rather than visualize nicely arced ovals for links, let's instead aim for simplicity. Links will be represented as lines in space. In order to do so, it will be essential to calculate the positions of each of the four nodes that connect three links in the body frame: (x_b, y_b). This function is termed `body_coords`, and returns a set of position pairs given links each of length $2L$ and two angles:

```
import numpy as np
import matplotlib.pyplot as plt

def body_coords(alpha,L):
    # Converts two body angles into the four positions of the three-link swimmer
    # in its own body frame
    dx =-np.cos(alpha[0])
    dy = np.sin(alpha[0])
    pos = np.zeros((4,2))
    pos[0,:] = 2*L*np.array([dx,dy])+np.array([-L, 0])
    pos[1,:] = np.array([-L,0])
    pos[2,:] = np.array([L,0])

    dx = np.cos(alpha[1])
    dy = np.sin(alpha[1])
    pos[3,:] = np.array([L,0])+2*L*np.array([dx, dy])

    return(pos)
```

Once you have this code working, try out different configurations and then plot the three-link swimmer, e.g.,

```
fig = plt.figure(figsize=(5,5))
ax = fig.gca()
pos = body_coords([np.pi/4, -np.pi/4],2)
plt.plot(pos[:,0],pos[:,1],'k',linewidth=3)
ax.axis('equal') # To fix the aspect ratio
```

Try out the code to make sure it does the right thing for different angles and for different link lengths. If it's working, it should look like the diagram below.

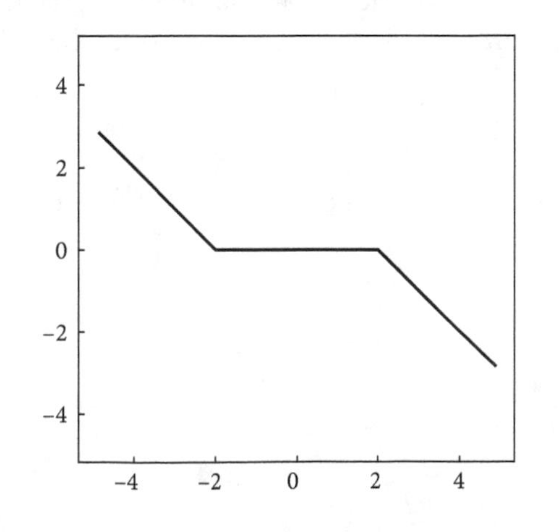

Now you're ready for the first challenge problem: to write a code that can generate the configuration space in which the three-link swimmer will operate.

CHALLENGE PROBLEM: Visualizing Configuration Space

Extend the code snippet to visualize a single body configuration as a configuration space from $-\pi/2 \leq \alpha_1 \leq \pi/2$ and $-\pi/2 \leq \alpha_2 \leq \pi/2$. If your code works, you should see an image that looks like the following:

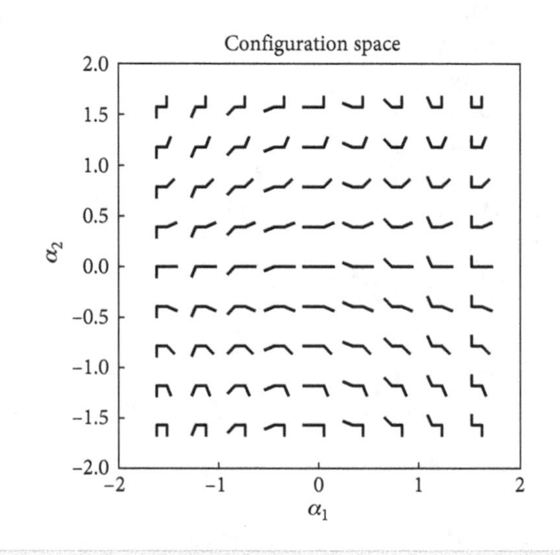

8.3 ORBITS IN CONFIGURATION SPACE

Before moving to the full Newtonian framework, it is essential to master the concept of orbits in configuration space, aka gaits. A gait is a repeated set of movements. For example, if the three-link swimmer were to move its links up and down, up and down, and so on, this can be thought of as a symmetric gait that traverses a line twice in the configuration space, albeit in opposite directions. Familiarizing yourself with the relationship between paths in configuration space and a representative sequence of configurations that make up a gait is essential to thinking about how organisms move in practice—whether for *E. coli*, spermatazoa, or sand lizards. To get you started, consider the following symmetric gait function:

```
def sym_gait(t,T):
    # Returns positions and velocities in angle space for customized gait in a
    # single Python list
    alpha1 = np.pi/4*np.cos(2*np.pi*t/T)
    alpha2 = np.pi/4*np.cos(2*np.pi*t/T)
    valpha1 = -2*np.pi/T*np.pi/4*np.sin(2*np.pi*t/T)
    valpha2 = -2*np.pi/T*np.pi/4*np.sin(2*np.pi*t/T)
    return [alpha1,alpha2,valpha1,valpha2]
```

This gait can be superimposed over the configuration space by adding the following code:

```
# Overlay the gait
gait_dyn = sym_gait(np.arange(0,1.01,0.01),1)
alpha1_gait = gait_dyn[0]
alpha2_gait = gait_dyn[1]
plt.plot(alpha1_gait,alpha2_gait,'r',linewidth=3)
```

This code yields a "gait." The gait is denoted as the light gray line in the configuration space below. In the left panel, that gait winds back over itself; that is, the gait starts with both α_1 and α_2 near 1 and then near -1 and then near 1 and so on. In the next challenge problem, it is your job to figure out a gait that can reproduce the sequence of configurations in the right panel.

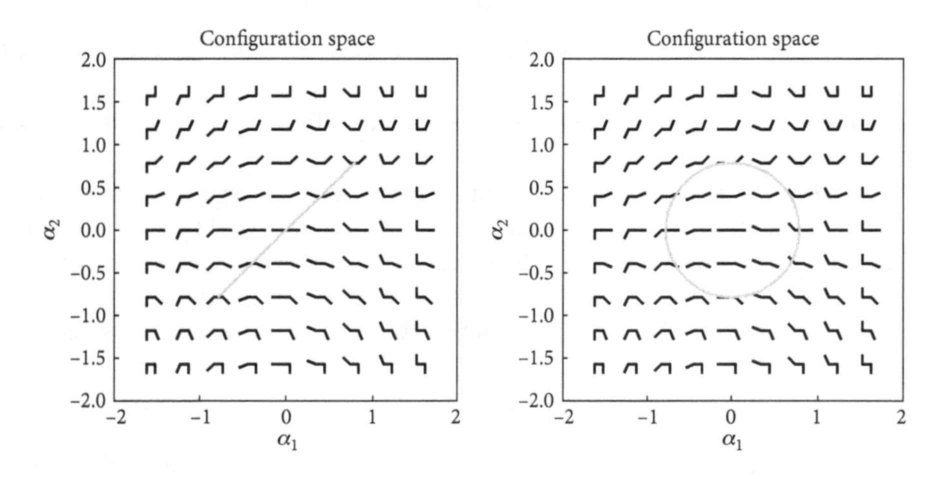

CHALLENGE PROBLEM: Designing Gaits

Develop an antisymmetric gait that reproduces the motions above and (optionally) a square-wave shape of an antisymmetric gait. If you have the time, write a code snippet that reproduces the configuration snapshots that appear below the panels in the figure.

8.4 FROM BORELLI TO NEWTON AND BACK AGAIN

8.4.1 Newtonian framework

How do we move from gait programs $(\alpha_1(t), \alpha_2(t))$ to dynamics in the world frame? The tack we take here is to focus on classical mechanics, i.e., $F = ma$, where F, m, and a denote force, mass, and acceleration, respectively. Conventionally, physics classes emphasize how simple physical interactions can be described in terms of second-order (usually linear) differential equations. However, such equations can be written in terms of a coupled set of first-order differential equations where the state space includes both position and velocity. That logic underlies the development of the FitzHugh-Nagumo model (introduced in the prior chapter). Here the same logic applies, albeit to the position (x, y), orientation θ, and corresponding velocities, i.e., translational velocities (v_x, v_y) and angular velocity ω_θ. Recognizing that positions change due to velocities and velocities change due to forces motivates the following set of equations for the dynamics of the three-link swimmer in the world frame:

$$\dot{x} = v_x \tag{8.1}$$

$$\dot{y} = v_y \tag{8.2}$$

$$\dot{\theta} = \omega_\theta \tag{8.3}$$

$$\dot{v}_x = \frac{F_x}{m} \tag{8.4}$$

$$\dot{v}_y = \frac{F_y}{m} \tag{8.5}$$

$$\dot{\omega}_\theta = \frac{\tau}{I} \tag{8.6}$$

Here F_x and F_y denote forces, m denotes the total mass, τ is the torque, and I is the moment of inertia.

These equations seem straightforward, but they also hide the key insight: that changes in the internal body configuration change the dynamics in the world frame. In essence, where are $\alpha_1(t)$ and $\alpha_2(t)$? The answer is embedded in a simple principle: for every action there is an equal and opposite reaction. Consider a particle moving with velocity v facing a drag k, such that the total force is $F = -kv$. In that event, one could write a simplified set of equations:

$$\dot{x} = v \tag{8.7}$$

$$\dot{v} = -kv/m \tag{8.8}$$

These equations are at the very core of understanding how the three-link swimmer interacts with its viscous environment (and, similarly, how the sandfish interacts with its dissipative, granular environment). In general, an organism moving with an initial velocity v_0 will coast for a while until it comes to a complete stop at position x_f. For the three-link swimmer, this coasting distance (and indeed the effective coasting time, τ) sets both length and time scales to compare to the length of the organism and the duration of the gait, T. In the limit that $x_f \ll L$ and $\tau \ll T$, then we should expect that, however the organism travels, it must be primarily due to internal body forces rather than due to inertia (and coasting). This insight underlies the next challenge problem.

CHALLENGE PROBLEM: Coasting Time and Distance

Consider a nonmotile organism of mass m with an initial velocity of v_0 in an environment with drag k. How far will the organism coast before it reaches a complete stop? What is the effective time over which the organism remains in motion? In answering this question, use a simulation and (if possible) mathematical analysis to verify your finding.

8.4.2 Borelli: Connecting external forces and internal body movements

In the limit that inertial forces are small and that coasting is minimal, then the internal movement of the body is the key to understanding how the swimmer self-propels. Indeed, it is drag that both slows the organism down and allows it to move; because, by moving a

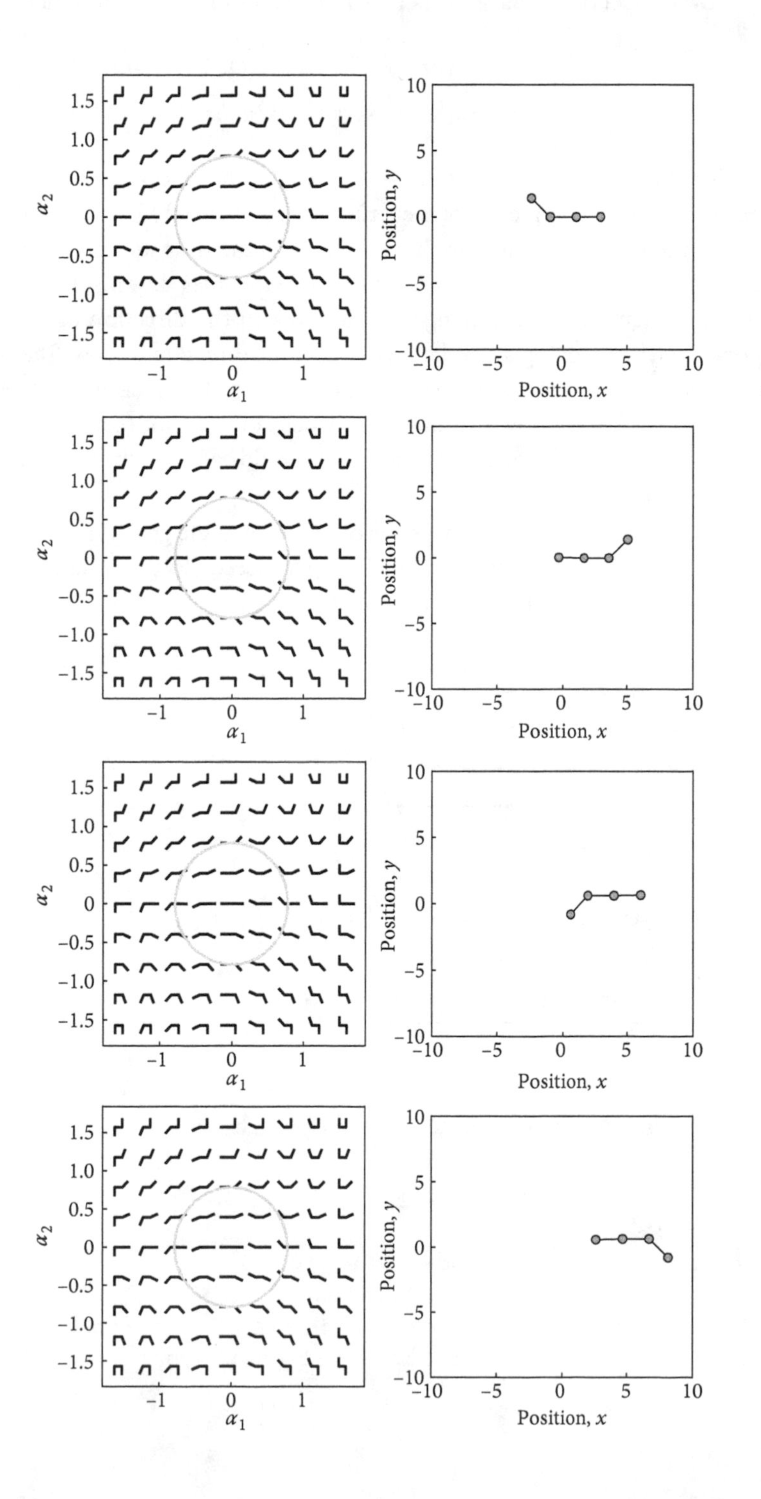

link in one direction, the organism must experience an equal force in the opposite direction. We have already assumed that $F_{drag} = -kv$. However, here we make the critical assumption that $F_{propulsion} = -kv_{link}$. In other words, if the link moves backward, then the force acting on the body is forward. Likewise, if the link moves forward, then the force acting on the body is backward. Hence, we must break down the forces into two components:

$$F_x = F_x(drag) + F_x(propulsion) \tag{8.9}$$

$$F_y = F_y(drag) + F_y(propulsion) \tag{8.10}$$

And there is yet one more complication—which is that the forces in the body frame of x and y must first be derived in terms of the forces acting on each link. Indeed, resistive force theory makes the critical assumption that one can add the drag and propulsion forces on each link to obtain the overall force. Further, the code includes the anisotropy with respect to the drag experienced when moving a slender link in the direction of flow versus the drag experience when moving a slender link against the direction of flow. The code itself is complicated, so you are encouraged to review the following two sections and perhaps even improve them! The code below is a minimal visualization script (see the book's website for more details). When the code is working, you should see something like the following images in an animated form.

The result should look like the images below. The complete gait (on the right) includes transient back- and forth-motion that nonetheless proceeds to the right, increasing in x, further assuming that angular velocity is fixed at 0.

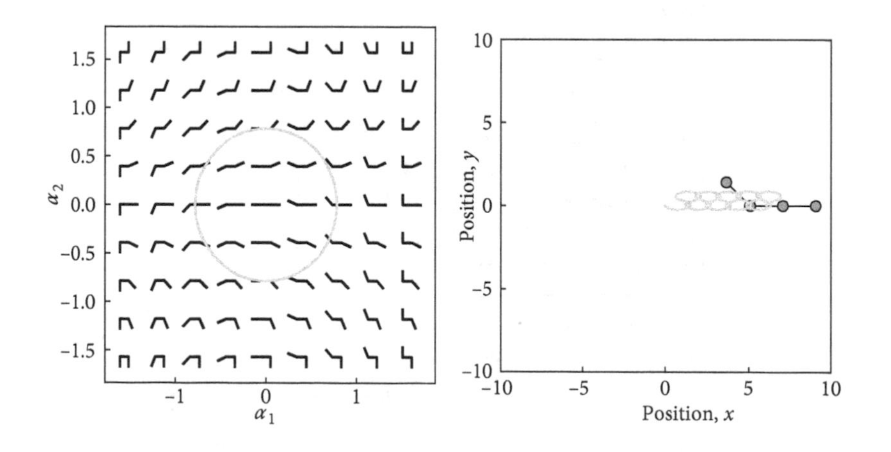

The full simulation code and master script are given below.

Complete simulation code

```
def model_4bead(y,t,pars):
    # RFT dynamics for a three-link swimmer
    # y -> bx, by, vbx, vby, vtheta, alpha1, alpha2
    # An eight-coordinate dynamic, only six of which are considered here
    # The other two are prescribed by the internal gait dynamics
```

```
# Coords
xb = y[0]
yb = y[1]
theta = y[2]
vxb = y[3]
vyb = y[4]
vtheta=y[5]
gait_dyn = pars['gait_function'](t,pars['T'])
alpha1 = gait_dyn[0]
alpha2 = gait_dyn[1]
valpha1 = gait_dyn[2]
valpha2 = gait_dyn[3]
L = pars['L']

# Point positions
# *----*----*----*
# 0    1    2    3


Amat = rotmat(theta+np.pi-alpha1)
dpos1 = np.matmul(Amat,np.array([2*L,0]))

Amat = rotmat(theta+alpha2)
dpos2 = np.matmul(Amat,np.array([2*L,0]))
bpos = np.array([[xb-L*np.cos(theta)+dpos1[0], yb-L*np.sin(theta)+dpos1[1]],
                   [xb-L*np.cos(theta), yb-L*np.sin(theta)],
                   [xb+L*np.cos(theta), yb+L*np.sin(theta)],
                  [xb+L*np.cos(theta)+dpos2[0], yb+L*np.sin(theta)+dpos2[1]]])
# Link vectors for anisotropy
link = np.zeros((4,2))
link[0,:] = bpos[0,:]-bpos[1,:]
link[1,:] = bpos[1,:]-bpos[2,:]
link[2,:] = bpos[2,:]-bpos[1,:]
link[3,:] = bpos[3,:]-bpos[2,:]

# Forces on each bead due to drag
bvel_vec = np.array([vxb, vyb])
fparr = np.zeros(np.shape(link))
fperp = np.zeros(np.shape(link))
for i in range(4):
    vec_projection = np.dot(bvel_vec, link[i,:])/(2*L)*link[i,:]/(2*L)
    vec_rejection = bvel_vec- np.dot(bvel_vec, link[i,:])/(2*L)**2*link[i,:]
    fparr[i,:] = -pars['kparr']*vec_projection
    fperp[i,:] = -pars['kperp']*vec_rejection

# Forces on the movable beads
fparr_bead = np.zeros((2,2))
fperp_bead = np.zeros((2,2))
```

```
    bead_vangle1 = vec_to_angle(link[0,0],link[0,1])-np.pi/2*np.sign(valpha1)
    bead_vvec1 = angle_to_vec(bead_vangle1)
    vec_projection = np.dot(bead_vvec1, link[0,:])/(2*L)*link[0,:]/(2*L)
    vec_rejection = bead_vvec1-np.dot(bead_vvec1, link[0,:])/(2*L)**2*link[0,:]
    fparr_bead[0,:] = -pars['kparr']*np.abs(valpha1)*2*L*vec_projection
    fperp_bead[0,:] = -pars['kperp']*np.abs(valpha1)*2*L*vec_rejection

    bead_vangle2 = vec_to_angle(link[3,0],link[3,1])+np.pi/2*np.sign(valpha2)
    bead_vvec2 = angle_to_vec(bead_vangle2)
    vec_projection = np.dot(bead_vvec2, link[3,:])/(2*L)*link[3,:]/(2*L)
    vec_rejection = bead_vvec2-np.dot(bead_vvec2, link[3,:])/(2*L)**2*link[3,:]
    fparr_bead[1,:]=-pars['kparr']*np.abs(valpha2)*2*L*vec_projection
    fperp_bead[1,:]=-pars['kperp']*np.abs(valpha2)*2*L*vec_rejection


    # Total force
    F = np.sum(fparr,axis=0)+np.sum(fperp,axis=0)+np.sum(fparr_bead,axis=0)+\
    np.sum(fperp_bead,axis=0)

    # Equations of motion
    dydt = np.zeros(6)
    dydt[0]=vxb
    dydt[1]=vyb
    dydt[2]=vtheta
    dydt[3]=pars['damping']*F[0]/pars['M']
    dydt[4]=pars['damping']*F[1]/pars['M']
    dydt[5]=0

    return dydt
```

Master script

```
pars={}
pars['L']=1
pars['kparr']=0.2
pars['kperp']=2
pars['max_angle']=np.pi/4
pars['damping']=1
pars['M']=0.1
pars['gait_function']= asym_gait
pars['T']=40
pars['cnt']=0

# Simulate
y0=np.zeros(6) # Six equations, all in world frame
t = np.arange(0,200.2,0.2)
y = integrate.odeint(model_4bead,y0,t,args=(pars,))
```

```
# Recover the gaits
alpha1 = np.zeros(np.shape(t))
alpha2 = np.zeros(np.shape(t))
valpha1 = np.zeros(np.shape(t))
valpha2 = np.zeros(np.shape(t))
for i in range(len(t)):
    gait_dyn = pars['gait_function'](t[i],pars['T'])
    alpha1[i]=gait_dyn[0]
    alpha2[i]=gait_dyn[1]
    valpha1[i]=gait_dyn[2]
    valpha2[i]=gait_dyn[3]

# Visualize swimming
# Do not forget to run the command %matplotlib qt first
fig, axes = plt.subplots(1,2,figsize=(10,5))

# Also show the shape space
ax=axes[0]
L = 0.05 # Artificially create ``small'' swimmers for visualization
dax = np.pi/8
day = np.pi/8
alpha1_vec = np.arange(-np.pi/2,np.pi/2+dax,dax)
alpha2_vec = np.arange(-np.pi/2,np.pi/2+day,day)
alpha1_grid, alpha2_grid = np.meshgrid(alpha1_vec, alpha2_vec)
for i in range(len(alpha1_vec)):
    for j in range(len(alpha2_vec)):
        pos = body_coords([alpha1_grid[j,i],alpha2_grid[j,i]],L)
        vis_pos = pos + np.array([alpha1_grid[j,i],alpha2_grid[j,i]])
        # Shift placement to alpha-1/alpha-2
        ax.plot(vis_pos[:,0],vis_pos[:,1],'k',linewidth=3)
ax.plot(alpha1,alpha2,c='r',linewidth=3)
plt.setp(ax.spines.values(),linewidth=2)
ax.tick_params(labelsize=15,direction='in',width=2)

ax = axes[1]
ax.set_xlim(-10,10)
ax.set_ylim(-10,10)
ax.set_aspect('equal')
plt.setp(ax.spines.values(),linewidth=2)
ax.tick_params(labelsize=15,direction='in',width=2)

# Go back to the swimming
for i in range(len(t)):
    ax.clear()
    ax.set_xlim(-10,10)
    ax.set_ylim(-10,10)
    ax.set_aspect('equal')
```

```
    pos = body_coords([alpha1[i],alpha2[i]],pars['L'])
    rpos = body_rotate(pos,y[i,2])
    vis_pos = np.array([y[i,0], y[i,1]])*np.ones(np.shape(rpos))+rpos
    ax.plot(vis_pos[:,0], vis_pos[:,1],'k',
            linestyle='-',
            marker='o',
            markeredgecolor='k',
            markerfacecolor=[0.5,0.5,0.5],
            markersize=8)

    plt.pause(0.001)
    ax.plot(y[:,0],y[:,1],c='r',linewidth=3)
    plt.show()
```

8.5 THE GREATEST GAIT OF ALL

Now that you have a working code, you should aim for the following:

- Try out the symmetric and antisymmetric gaits. Which of them leads to translation and why?
- By changing the magnitude of angular change, can you modify the horizontal translation per gait cycle? Is there an optimal angle for the antisymmetric gait?
- How does the mass change the ability of the three-link swimmer to self-propel?
- How much does anisotropy matter? That is, if one sets the parallel and perpendicular drag forces equal, does that decrease or increase the amount of translation per gait cycle?

Finally, if you have time, see if you can outdo the antisymmetric gait and measure the translation per gait cycle. If you were the three-link swimmer, how would you want to move? These targets form the basis for the homework and are central to the development of theory, simulations, analysis of living systems, and the construction of real-world robots that leverage "geometric mechanics."

SOLUTIONS TO CHALLENGE PROBLEMS

SOLUTION: Visualizing Configuration Space

There are many ways to solve this problem. This solution set uses a vector of values, generates a "mesh" of (α_1, α_2), and then finds the body coordinates for each set of angles. It is up to you to decide how finely to divide the configuration space; here the space includes 9 values of α_1 and α_2, for 81 potentially different configurations:

```
fig = plt.figure()
ax =  fig.gca()
L = 0.05 # Artificially create ''small'' swimmers for visualization
dax = np.pi/8
```

```
day = np.pi/8
alpha1_vec = np.arange(-np.pi/2,np.pi/2+dax,dax)
alpha2_vec = np.arange(-np.pi/2,np.pi/2+day,day)
alpha1, alpha2 = np.meshgrid(alpha1_vec, alpha2_vec)

for i in range(len(alpha1_vec)):
    for j in range(len(alpha2_vec)):
       pos = body_coords([alpha1[j,i],alpha2[j,i]],L) # Configuration
       vis_pos = pos + np.array([alpha1[j,i],alpha2[j,i]])
       plt.plot(vis_pos[:,0],vis_pos[:,1],'k',linewidth=3)
plt.xlabel(r'$\alpha_1$',fontsize=20)
plt.ylabel(r'$\alpha_2$',fontsize=20)
plt.title('Configuration space',fontsize=20)

plt.xlim(-2,2)
plt.ylim(-2,2)
ax.set_aspect('equal','box')
plt.setp(ax.spines.values(),linewidth=2)
ax.tick_params(labelsize=15,direction='in',width=2)
```

SOLUTION: Designing Gaits

This solution has two components. First is the asymmetric gait, which can be coded as follows:

```
def asym_gait(t,T):
    # Returns positions and velocities in angle space for customized gait in a
    # single Python list
    alpha1 = np.pi/4*np.cos(2*np.pi*t/T)
    alpha2 = np.pi/4*np.sin(2*np.pi*t/T)
    valpha1 = -2*np.pi/T*np.pi/4*np.sin(2*np.pi*t/T)
    valpha2 = 2*np.pi/T*np.pi/4*np.cos(2*np.pi*t/T)
    return [alpha1,alpha2,valpha1,valpha2]
```

Next, a gait sequence can be plotted using the following code snippet, which turns off the axis to highlight the internal configuration changes.

```
fig = plt.figure()
ax =  fig.gca()

for t in np.arange(0,1.125,0.125):
    gait_dyn = sym_gait(t,1)
    alpha1 = gait_dyn[0]
    alpha2 = gait_dyn[1]
    pos = body_coords([alpha1,alpha2],0.015)
```

```
    vis_pos = pos + np.array([t, 0])
    plt.plot(vis_pos[:,0],vis_pos[:,1],'k',linewidth=3)
plt.xlim(-0.1,1.1)
plt.ylim(-0.2,0.2)
plt.axis('off')
```

SOLUTION: Coasting Time and Distance

This coupled set of linear differential equations can be solved, first by recognizing that $v(t) = v_0 e^{-kt/m}$. As a result, $\dot{x} = v_0 e^{-kt/m}$, which implies that

$$dx = v_0 e^{-kt/m} dt. \tag{8.11}$$

Integrating both sides yields

$$x(t) = -\frac{v_0 m}{k} \left(e^{-kt/m} \right) \Big|_0^\infty \tag{8.12}$$

such that $x(t) = \frac{v_0 m}{k}$. Hence, larger organisms will coast longer; likewise, less viscous environments (with less drag) will also lead to longer coasting distances. Note that the effective time scale of coasting is $\tau = m/k$. The following code simulates the model and verifies the relationship shown in the figure below (either for variation in m or in k).

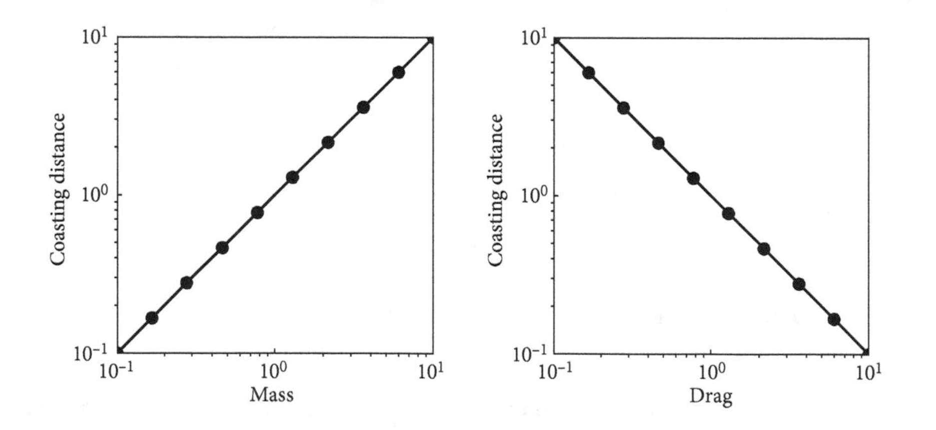

The *key takeaway* here is that the ratio of m to k sets a time scale (and therefore a coasting distance) over which inertia can act. When $(m/k) \ll T$, the gait period, then it must be internal forces, rather than inertial forces, that control propulsion.

Dynamic code

```
def model_coast(y,t,pars):
    dydt = np.zeros(2)
```

```
    dydt[0] = y[1]
    dydt[1] = -pars['k']*y[1]/pars['m']
    return dydt
```

Simulation and visualization code

```
# Simulation
mrange = np.logspace(-1,1,num=10)
v0=1
pars={}
yf = np.zeros(np.shape(mrange))
for i in range(len(mrange)):
    pars['m']=mrange[i]
    pars['k']=1
    t = np.linspace(0,100)
    y0 = np.array([0,v0])
    y = integrate.odeint(model_coast,y0,t,args=(pars,))
    yf[i] = y[-1,0]

# Plot simulation against theory
fig = plt.figure()
ax = fig.gca()
m = np.logspace(-1,1,num=1000)
plt.loglog(m,m*v0/pars['k'],'k',linewidth=3)
plt.loglog(mrange,yf,'k',
           linestyle='',
           marker = 'o',
           markersize = 10)
plt.xlabel('Mass',fontsize=20)
plt.ylabel('Coasting distance',fontsize=20)
```

Part III

Populations and Ecological Communities

Chapter Nine

Flocking and Collective Behavior: When Many Become One

9.1 AGENT-BASED MODELS AND EMERGENCE IN FLOCKS

Birds of a feather flock together. This old adage reminds us of certain apparent truisms—both good and bad—that characterize communities. There are advantages to sticking together, whether the groups are insects, fish, or primates. These advantages arise from prosocial behaviors, including defense, feeding, mating, and yes, even play. Yet organisms in motion don't just have to stick together, like birds on a wire in the words of Leonard Cohen's classic folk tune; they also have to move together. Moving together is a form of emergent choreography. And, until the mid-1990s, it was unclear exactly how organisms could accomplish such a "group"-level trait. It was generally understood that when a group of starling pigeons moves, nearly fluidlike, in the evening sky, there is not a gene in each that encodes that particular set of movements, at that particular time and place, and specifies each bird's particular position in the flock. That seems obvious. But, if such flock movements are not prescribed, then it points to something challenging: what are the local rules of interaction that govern movement and how can local rules lead to emergent population-level dynamics?

Answering this question forms the basis for this laboratory, which explores the Vicsek model of flocking behavior. The Vicsek model, or self-propelled particle (SPP) model, was proposed in the mid-1990s (Vicsek et al. 1995). The core idea is revealed in the following figure:

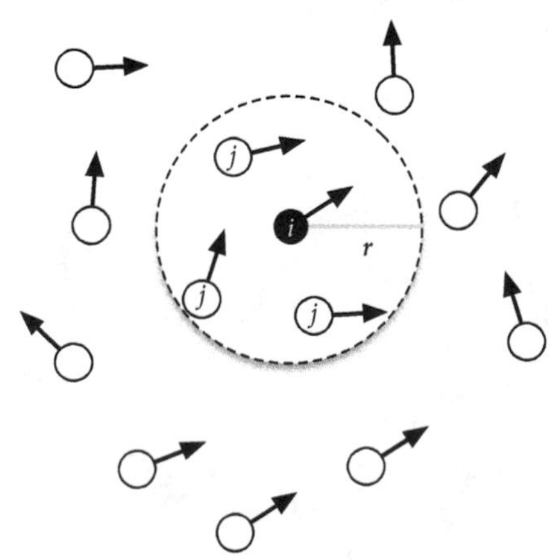

The image depicts a focal individual (black circle) surrounded by other individuals (open circles); all of the individuals are moving, but not necessarily in the same direction. The focal individual, i, senses the movements of organisms close to it. In this example, the term "close" applies to all those individuals j whose current position is within r of individual i. The model then implements a rule: that the direction in which a focal organism moves is a function of its current velocity, v_i, and the velocity of its close neighbors, $\{v_j\}$, as well as some element of randomness. If this same rule applies to all particles, then it is possible for this local rule to translate into including emergent dynamics seemingly quite unexpected and unrelated to the local rule.

This lab provides a basis for moving from local rules, as described by SPP theory, but you will be expected to implement the entire code. The code then becomes the basis for exploring dynamics as parameters vary, with a particular emphasis on the transition from individual-like behavior to flock-like behavior. As it turns out, this transition has all the properties of phase transitions in the study of physical materials (water to ice) and in the study of static, complex systems (as in percolation theory, or connectivity of some and then suddenly all agents in a system). Here a notable difference is that there is a dynamic order parameter associated with very long length scales that approach that of the system size in

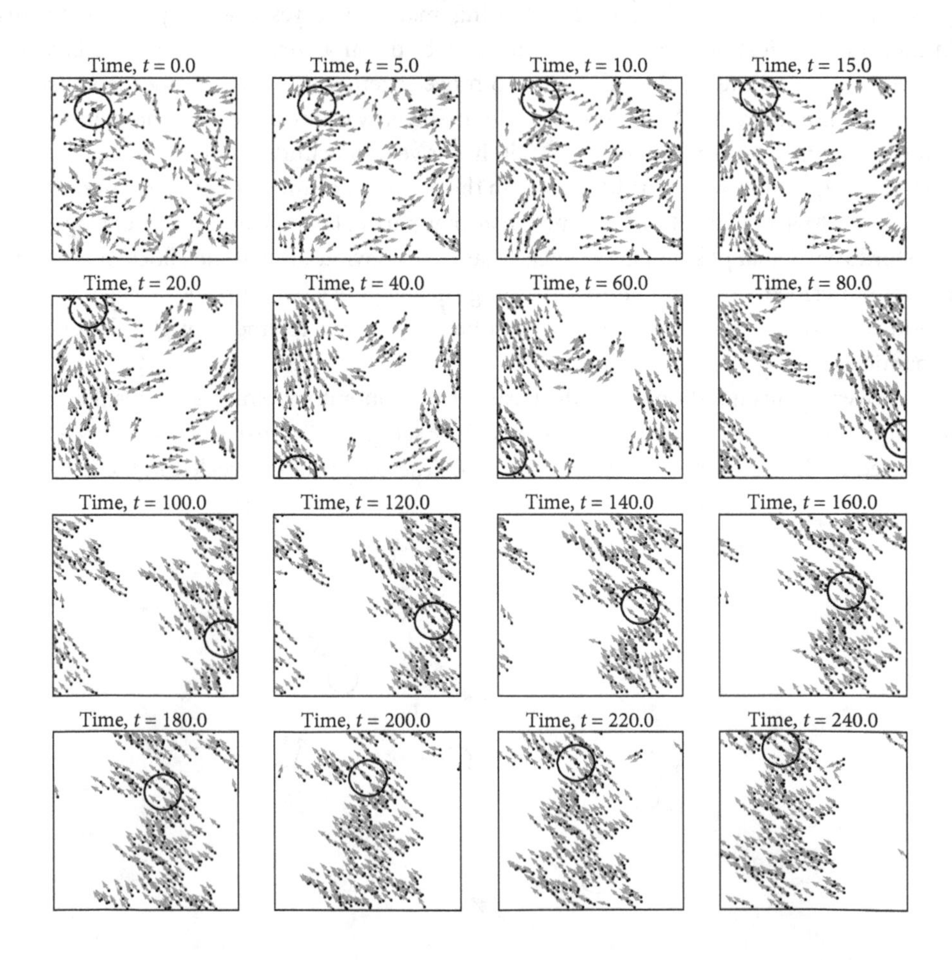

which the behavior appears coherent and (nearly) identical. The snapshots illustrate this point. Initially, a collection of 300 individuals moves at random (at $t = 0$). Given local feedback, these individuals spontaneously start to move together in flocks, sometimes spanning nearly the entire population. These flocks emerge even if no one organism has any preference regarding where or why to move in a particular direction. As a guide to the eye, pay attention to the focal individual, arbitrarily surrounded by a black circle—even if it is not moving with the flock at the start, it soon is, along with all the other individuals. This form of spontaneous symmetry breaking and emergent order is the centerpiece of this computational laboratory.

9.2 THE VICSEK MODEL

The Vicsek or SPP model simulates the dynamics of flocking organisms, or *boids*. The emergence of flocking requires that local information is transmitted to groups and potentially to the entire population. This transmission is not encoded directly in the model, but is a potential outcome given suitable conditions. At each time point, the direction of motion for each boid is determined from both the average direction of neighboring boids within a radius, $r = 1$, and a noise term, η. We follow the original paper (Vicsek et al. 1995) and implement the model in a 2D setting with periodic boundary conditions. The pseudocode for the SPP model is as follows:

- Initialize the position $(x_i(t), y_i(t))$ and direction $\theta_i(t)$ of each boid—note the model maintains a constant speed of each boid while it changes its direction.
- Update the direction of every boid's movement, $\theta_i(t+1)$, by reorienting based on the local direction of neighbors and an additional noise term:

$$\theta_i(t+1) = \arctan\frac{\langle \sin\theta_j(t)\rangle_R}{\langle \cos\theta_j(t)\rangle_R} + \pi\Theta(-\langle \cos\theta_j(t)\rangle_R) + noise$$

- Update the position of each boid i assuming ballistic motion:

$$x_i(t+1) = x_i(t) + v_0\Delta t\cos(\theta_i(t+1))$$
$$y_i(t+1) = y_i(t) + v_0\Delta t\sin(\theta_i(t+1))$$

That is the entirety of the model. However, the challenge lies in the updating component and represents the first challenge problem.

CHALLENGE PROBLEM: Reorienting Boids

Write a code to reorient boids based on the orientation of neighbors, i.e., including both an averaging component and a random term. (Hint: Identify a solution to the problem that arises in defining a local "group" if the focal boid is alone.)

9.2.1 Numerical implementation

When fully integrated, the main code tracks the x and y positions of each boid along with the orientation θ of motion for a set number of time steps, T. The simulation begins by randomly placing each boid in the domain. Additionally, the initial velocity of each boid is oriented in a random direction θ between $[0, 2\pi]$. It is now your turn to build your SPP model.

CHALLENGE PROBLEM: SPP Simulation Code

Set up the main code to initialize the dynamics and loop through time. A skeleton for the code is as follows. The input parameters to this function are the total number of boids, N, the linear dimension of the domain, L, and the amount of noise, η. (Hint: Implement the dynamics with periodic boundary conditions, but keep this in mind while trying to detect which boids are close to each other.)

```
def boids_dyn(pars):
    # function [xt,yt,thetat]=boids_dyn(N,L,eta)
    # Inputs
    # Inputs include N - number of boids,
    # L - linear dimension, eta - noise strength
    # Outputs
    # xt and yt -- positions of each boid over time
    # thetat -- direction of velocity of each boid over time
    # Simulation parameters
    totsteps = 1000
    delt = 1
    v0 = 0.03
    R = 1

    # Initial conditions
    currx = L*np.random.uniform(size=N)
    curry = L*np.random.uniform(size=N)
    currtheta = 2*np.pi*np.random.uniform(size=N)

    # Preallocate output
    xt = np.zeros((totsteps,N))
    yt = np.zeros((totsteps,N))
    thetat = np.zeros((totsteps,N))

    # Run the sim (fill this in)
    # Loop t through time steps

    # Loop i through each boid

    # Find the boids within R of i

    # Calculate the new direction of boid i
    # based on the directions of neighbors
```

```
    # End Loop i

    return [xt,yt,thetat]
```

9.2.2 Visualization basics

The visualization of boids can be accomplished in multiple ways. However, to represent both the position and velocity of boids, it can be useful to implement a visualization that goes beyond merely showing where the organisms are to visualizing where they are going. For example, if the variables `xt`, `yt`, `vxt`, `vyt` represent x, y, v_x, and v_y, respectively, it is possible to use the following command for visualization:

```
plt.quiver(xt,yt,vxt,vyt,color='grey')
plt.plot(xt,yt,'k.',linestyle='')
```

The `quiver` command can be modulated with optional parameters, and the second command places a circle where the boid is. Here is a snapshot of example dynamics. Notice how the direction of boid movement can be different even if the boids are near each other. The high noise causes deviations from the ordering.

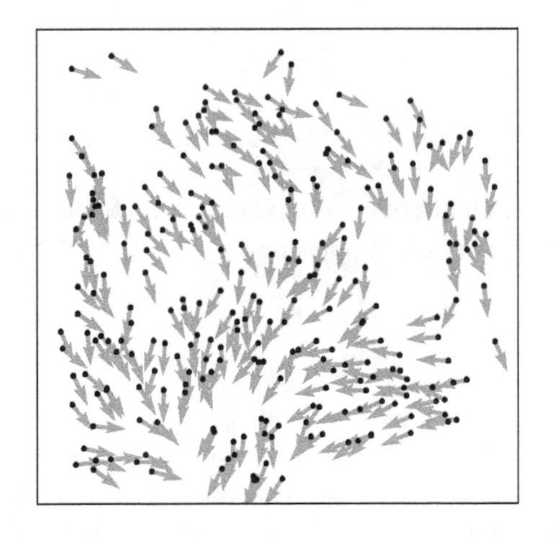

At this point, you are ready to explore the transition from individual to collective motion in the SPP model framework.

9.3 FLOCKING DYNAMICS

9.3.1 Baseline case

The dynamics of boids in the SPP model can be recapitulated in two regimes: the low- and the high-noise regimes. Here we graphically explore the dynamics of the Vicsek model. First, we simply explore a series of images taken from the initial configuration to $t = 1200$ in a boid simulation with the following parameters: $N = 250$, $L = 5$, corresponding to a density of 10 boids per unit square, $v_0 = 0.03$, $dt = 0.2$. Next, we explore the low-noise case where

$r = 0.5$ and $\eta = 0.1$ and the high-noise case where $r = 0.05$ and $\eta = 1$. The outcomes are stark. Hence, the first challenge problem is to recapitulate the dynamics and develop a plotting code for these SPP flipbooks.

CHALLENGE PROBLEM: Visualizing the Flock

Develop a visualization code that generates a panel of 16 images given a specified time set and positions and velocities of each boid. Use this code to visualize the low- (left) and high-noise (right) cases. If your code works, it should look (nearly) like this:

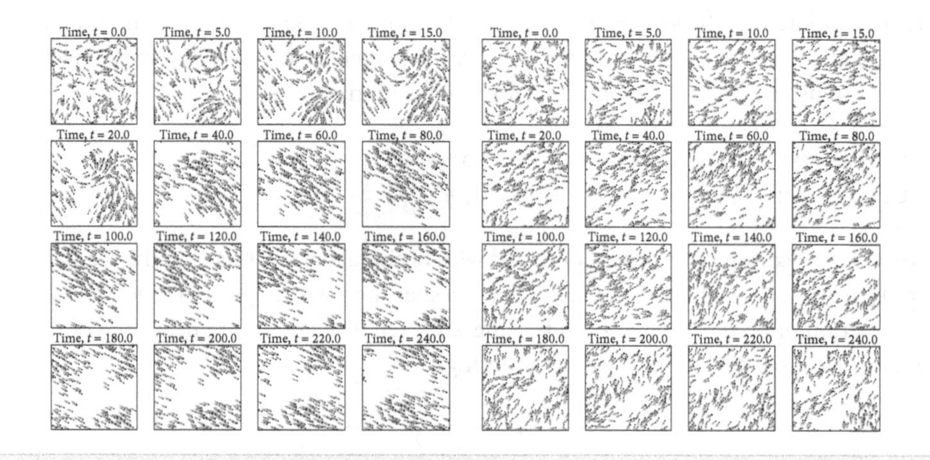

CHALLENGE PROBLEM: Flock of Boids—the Movie

Building on the prior code, develop a movie version of your flock in a single panel, then watch it. It will be particularly appealing if you use $\eta = 0.25$, $r = 0.5$, and $N = 500$.

9.3.2 Varying noise

The parameter η controls the amount of noise in the model. As $\eta \to \pi$, each boid essentially moves in a random direction. Increasing noise reduces the correlation of angles between neighboring boids. Quantifying the ordering requires an *order parameter*, i.e., a feature that describes the macroscopic behavior of the system rather than the behavior of an individual or small set of individuals. However, when many behave like one, then the order parameter can change rapidly from 0 (disordered, individuals acting distinctly) to 1 (ordered, individuals acting as a flock). The order parameter we focus on is the average velocity in the SPP model:

$$\bar{v} = \frac{1}{N v_0} \left| \sum_i^N \vec{v}_i \right| .$$

The average velocity is equal to the sum of velocity vectors divided by the sum of the magnitude of each vector. An order parameter near 0 means the boids are not correlated in direction, and an order parameter near 1 means all the boids face nearly the same direction. The following function calculates the order parameter:

```
def dyn2orderpar(thetat):
# Calculate the average velocity - order parameter
    N=len(thetat[0,:])
    vxt=np.cos(thetat)
    vyt=np.sin(thetat)
    sumvxt=np.sum(vxt,axis=1)
    sumvyt=np.sum(vyt,axis=1)
    normvt=np.sqrt((sumvxt)**2+(sumvyt)**2)/N
    return normvt
```

The following graph shows the dynamics of $\bar{v}$ for both a disordered state and an ordered state using $N=250$, $L=5$, $r=0.3$, $\eta=0.25$ (for the ordered state) and $\eta=2.5$ (for the disordered state) as simulated over 1000 time steps:

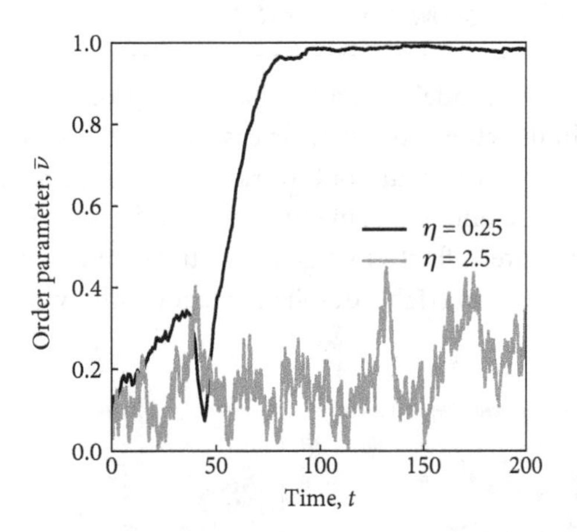

This illustration provides a sense that η can act to modify the degree of order in the system, in concert with other parameters. It is worthwhile to explore the dynamics of these systems using the movie code. The differences are stark, and range from flocking to seemingly senseless flickering in direction.

CHALLENGE PROBLEM: Coherence and Critical Levels of Noise

At what level of noise η does the final value of the order parameter drop nearly to 0? In doing so, use a density, e.g., $N=300$, $r=0.3$, and a range of η from 0 to 5.

When phase transitions occur, order parameters follow a scaling that depends on the microscopic dynamics. Vicsek et al. (1995) estimate the critical noise value as $\eta_c=2.9$. This is the noise level below which the boids begin to flock based on the order parameter, $\bar{v}$. If you have the time, rescale the previous plot and plot with logarithmic axes to show the power law behavior of the order parameter near the critical point. That is, see if the data you have follows

$$v_\alpha \sim (\eta_c - \eta)^\beta$$

where β shows the scaling of the order parameter near the phase transition.

9.4 BONUS: THE POWER OF LEADERSHIP

If you are looking for a challenge, try to designate a few "informed" boids who have a preferred direction and see how many it takes to get the others to follow their lead. For example, consider the case of a fixed set m of boids, $i = 1 \ldots m$, who all have the same preference for their random direction. This bonus challenge requires that you modify your core code to include informed individuals who always want to go in their preferred direction, albeit with noise.

CHALLENGE PROBLEM: Follow the Leader(s)

Develop a modified SPP model with a flexible number of informed individuals with fixed preferences for direction. As a specific case, try $m = 5$ informed individuals each with a preference of $\theta = 0$. If your code works, you should be able to use the same random seed and see the differences in outcome; e.g., for $r = 0.5$, $\eta = 0.25$, and $N = 500$, the results contrast between flocks that go in arbitrary directions (left) and those that go in the preferred direction (right) despite the fact that only 1 in 100 individuals has a preference!

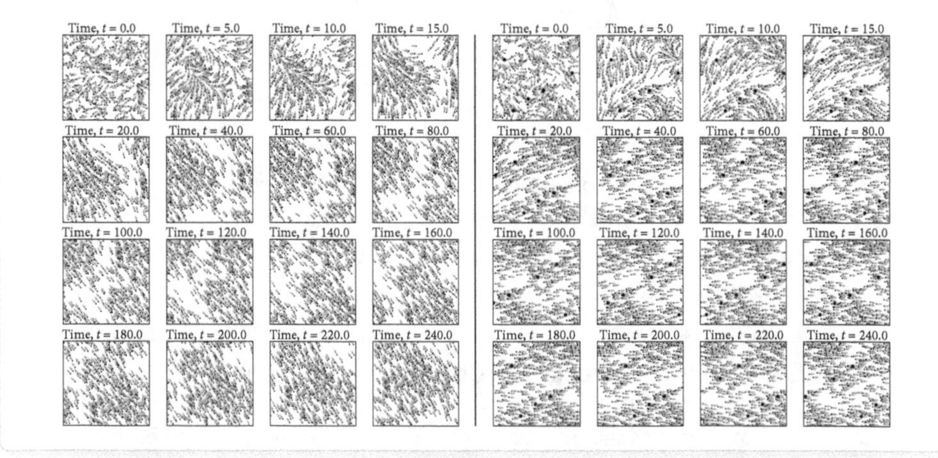

This challenge also directly relates to models in which individuals balance their personal preference and the local group preference. Such models can be tuned across a range of weightings and subsets, including leading to unusual outcomes where too many informed leaders leads to indecision such that the group cannot converge on a single direction. For those who cook, the old adage "too many cooks in the kitchens" comes to mind. The homework develops these ideas further, helping you link computational models to the emergent behavior of flocks and groups.

SOLUTIONS TO CHALLENGE PROBLEMS

SOLUTION: Reorienting Boids

In the SPP model, the direction is updated based on the orientation of neighboring boids and noise:

$$\theta_i(t+1) = \arctan\frac{\langle \sin\theta_j(t)\rangle_R}{\langle \cos\theta_j(t)\rangle_R} + \pi\Theta(-\langle \cos\theta_j(t)\rangle_R) + noise$$

where the averages $\langle * \rangle_R$ are over the indices of j corresponding to the boids within R of the focal boid indexed by i. The numerator in the argument of the arctan is the average y direction of the neighboring boid velocities, and the denominator is the average x direction of the neighboring boid velocities. By taking the arctan of the ratio, we get a new angle. However, the range of the arctan operator is between $[\frac{-\pi}{2}, \frac{\pi}{2}]$, thus if the velocity in the x direction is less than 0, we must correct by shifting by π. This is the second term, where we utilize the Heaviside function, $\Theta(x)$, which is 1 if $x > 0$ and 0 otherwise (note the minus sign in the argument of the Heaviside function in the main equation). Consider different combinations of the signs of the x and y components of the velocity and determine the appropriate correction term. Last, the noise term is uniformly sampled random numbers between $[-\frac{\eta}{2}, \frac{\eta}{2}]$.

This model can be implemented as follows, assuming that `currx` and `curry` denote the positions of all boids and `currtheta` denotes the current angle. In this code, the noise is scaled between $-\eta$ and η, and this enables modulation to favor local averaging. The key idea is to calculate the local distances using periodic boundary conditions (hence the use of the `min` distance metric).

```
# Loop i through each boid
for i in range(N):
    # Find distances
    dx = np.minimum(np.abs(currx-currx[i]),
                    np.abs(currx-(currx[i]+L)),
                    np.abs(currx-(currx[i]-L)))
    dy = np.minimum(np.abs(curry-curry[i]),
                    np.abs(curry-(curry[i]+L)),
                    np.abs(curry-(curry[i]-L)))
    boid_r2 = np.sqrt( (dx)**2 + dy**2 )
    # Identify the set of local boids
    avgthetas = currtheta[boid_r2<=r]
    # Find the orientation of local boids
    avgx = np.mean(np.cos(avgthetas))
    avgy = np.mean(np.sin(avgthetas))
    # Reset the new direction
    newtheta[i] = np.arctan2(avgy,avgx)
deltheta = -eta/2+eta*np.random.uniform(size=N)
currtheta = newtheta+deltheta
```

SOLUTION: SPP Simulation Code

The final code is below. A critical set of lines is how to find those particles within r of the focal boid. A “naive” choice of squared distances does not work in the periodic boundary conditions. Hence, one must first find the minimum distance in both the x and y directions before selecting the subset. The other challenge is to reposition boids in the domain between 0 and L if they move outside. This simulation assumes that the boids are moving in a 2D domain with periodic boundary conditions to reduce finite size effects. However, alternatives can also be imposed such that there are hard boundary conditions.

```
def boids_dyn(pars):
    # function [xt,yt,thetat]=boids_dyn(N,L,eta)
    # Inputs
    # Inputs include N - number of boids,
    # L - linear dimension, eta - noise strength

    # Outputs
    # xt and yt -- positions of each boid over time
    # thetat -- direction of velocity of each boid over time

    # Simulation parameters
    N=pars['N']
    eta=pars['eta']
    L=pars['L']
    totsteps=pars['totsteps']
    delt=pars['dt']
    v0=pars['v0']
    r=pars['r']

    # Initial conditions
    currx = L*np.random.uniform(size=N)
    curry = L*np.random.uniform(size=N)
    currtheta = 2*np.pi*np.random.uniform(size=N)

    # Preallocate output
    xt = np.zeros((totsteps,N))
    yt = np.zeros((totsteps,N))
    thetat = np.zeros((totsteps,N))

    # Run the sim (fill this in)
    # Loop t through time steps
    for t in range(totsteps):
        # Preallocate new directions
        newtheta = np.zeros(N)
```

```
        # Loop i through each boid and determine new directions
        for i in range(N):
            # find the boids within R of i
            dx = np.minimum(np.abs(currx-currx[i]),
                            np.abs(currx-(currx[i]+L)),
                            np.abs(currx-(currx[i]-L)))
            dy = np.minimum(np.abs(curry-curry[i]),
                            np.abs(curry-(curry[i]+L)),
                            np.abs(curry-(curry[i]-L)))
            boid_r2 = np.sqrt( (dx)**2 + dy**2 )
            # Calculate the new direction of boid i
            # based on the directions of neighbors
            avgthetas = currtheta[boid_r2<=r]
            # Average x and y components of vectors
            avgx = np.mean(np.cos(avgthetas))
            avgy = np.mean(np.sin(avgthetas))
            # Obtain average direction from arctan
            newtheta[i] = np.arctan2(avgy,avgx)
            # end Loop i
        # Obtain vector of noise to apply to boid directions
        deltheta = -eta/2+eta*np.random.uniform(size=N)
        # Update directions
        currtheta = newtheta+deltheta
        # Update positions based on ballistic motion and new directions
        currx = np.mod(currx + v0*np.cos(currtheta)*delt, L)
        curry = np.mod(curry + v0*np.sin(currtheta)*delt, L)
        # Track the updates over time
        xt[t,:] = currx
        yt[t,:] = curry
        thetat[t,:] = currtheta

    return [xt,yt,thetat]
```

SOLUTION: Visualizing the Flock

The visualization of a dynamic flock requires conveying a sense of time. The most direct way is to use a movie, i.e., by capturing multiple stills and then visualizing them together. But this is a book, so connected still images can also serve an important purpose. This is the solution presented in the code below; the choice of a 4×4 grid is arbitrary. It is also possible to construct your own axes, but this takes more time. To switch from flipbooks to video, simply focus on one plot and replace the quivers so you can see a flock in action.

```
nstrip = 16
tmpt=np.concatenate(([0,25,50,75],np.arange(100,1201,100)))
fig = plt.figure(figsize=(24,24))
for i in range(nstrip):
    ax = plt.subplot(4,4,i+1)
    plt.plot(xt[tmpt[i],:],yt[tmpt[i],:],'k.',linestyle='')
    plt.quiver(xt[tmpt[i],:],yt[tmpt[i],:],
               np.cos(thetat[tmpt[i],:]),np.sin(thetat[tmpt[i],:]),
               color='grey',scale=15,width=0.008)
    plt.title('Time, $t={time}$'.format(time=tmpt[i]*pars['dt']))
    plt.xlim([0,pars['L']])
    plt.ylim([0,pars['L']])
    plt.yticks([])
    plt.xticks([])
```

SOLUTION: Flock of Boids—the Movie

The function `movie_dyn` uses the same plot commands as above and takes as input the position and angles to plot each of the boids as they move in time and space. This is not a complicated code, but it is helpful and can allow you to watch the collective in a continuous, aka movie, mode. The `pause` command is useful for setting the effective frame rate.

```
def movie_dyn(xt,yt,thetat,pars):
    fig = plt.figure(figsize=(5,5))
    ax = fig.gca()
    for i in range(len(xt)):
        ax.clear()
        plt.plot(xt[i,:],yt[i,:],'k.',linestyle='')
        plt.quiver(xt[i,:],yt[i,:],
                   np.cos(thetat[i,:]),np.sin(thetat[i,:]),
                   color='grey',scale=15,width=0.008)
        plt.xlim(0,pars['L'])
        plt.ylim(0,pars['L'])
        plt.xticks([])
        plt.yticks([])
        plt.pause(0.01)
    return fig
```

SOLUTION: Coherence and Critical Levels of Noise

Suppose we vary the strength of the noise, η, between 0 and 5. Run the dynamics for 1000 time steps and report the order parameter averaged over the next 1000 time steps. The result is as follows:

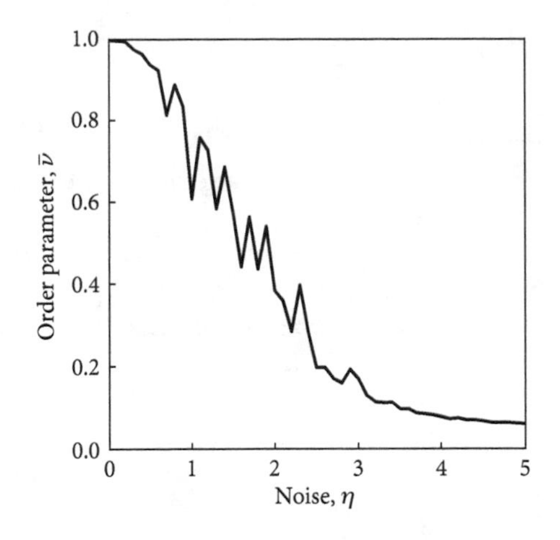

Part of the challenge here is that the order parameter can fluctuate as collective flocks emerge and come apart. Visually, we see that the order parameter begins to drop nearly to 0 when $\eta \approx 3$.

SOLUTION: Follow the Leader(s)

The following code includes a new component to the dynamics: individuals with fixed preferences on where to go. In this case, there are two new variables: m, the number of informed individuals, and $\{\theta_m\}$, the set of informed directions. Many variants are possible. Here the model assumes the preferences are fixed, subject to noise. As is apparent in the challenge problem, although flocking occurs without informed leaders, a very small number of leaders can "steer" the entire flock reliably to their preferred direction.

```
def boids_dyn_leader(pars):
    N=pars['N']
    eta=pars['eta']
    L=pars['L']
    totsteps=pars['totsteps']
    delt=pars['dt']
    v0=pars['v0']
    r=pars['r']
    thetam=pars['thetam']
    m=pars['m']
    # Initial conditions
    currx = L*np.random.uniform(size=N)
    curry = L*np.random.uniform(size=N)
    currtheta = 2*np.pi*np.random.uniform(size=N)

    # Run the sim
    xt = np.zeros((totsteps,N))
    yt = np.zeros((totsteps,N))
    thetat = np.zeros((totsteps,N))
```

```
for t in range(totsteps):
    # Preallocate new directions
    newtheta = np.zeros(N)
    newtheta[:m+1]=thetam
    # Loop i through each boid and determine new directions
    for i in np.arange(m+1,N):
        # Find the boids within R of i
        dx = np.minimum(np.abs(currx-currx[i]),
                        np.abs(currx-(currx[i]+L)),
                        np.abs(currx-(currx[i]-L)))
        dy = np.minimum(np.abs(curry-curry[i]),
                        np.abs(curry-(curry[i]+L)),
                        np.abs(curry-(curry[i]-L)))
        boid_r2 = np.sqrt( (dx)**2 + dy**2 )
        # Calculate the new direction of boid i
        # based on the directions of neighbors
        avgthetas = currtheta[boid_r2<=r]
        # Average x and y components of vectors
        avgx = np.mean(np.cos(avgthetas))
        avgy = np.mean(np.sin(avgthetas))
        # Obtain average direction from arctan
        newtheta[i] = np.arctan2(avgy,avgx)
        # end Loop i
    # Obtain vector of noise to apply to boid directions
    deltheta = -eta/2+eta*np.random.uniform(size=N)
    # Update directions
    currtheta = newtheta+deltheta
    # Update positions based on ballistic motion and new directions
    currx = np.mod(currx + v0*np.cos(currtheta)*delt, L)
    curry = np.mod(curry + v0*np.sin(currtheta)*delt, L)

        # Track the updates over time
        xt[t,:] = currx
        yt[t,:] = curry
        thetat[t,:] = currtheta

            return [xt,yt,thetat]
```

Chapter Ten

Conflict and Cooperation Among Individuals and Populations

10.1 STRATEGIES, GAMES, AND POPULATIONS

This lab explores the link between strategies, games, and population outcomes. The core ideas underlying this laboratory are depicted in Figure 10.1. Here hawks and doves compete for nesting sites. When two hawks interact they fight, so there can be a strong cost for such aggressive behavior. In contrast, when two doves interact, they share the nesting site. Finally, when a dove and a hawk meet, the dove flies away and the hawk captures the nesting site. These simplified interactions can be framed in the language of game theory. Benefits accrued to different types of organisms (or strategies) then influence the relative abundance of those individuals. As the relative abundance changes, then the types of each kind of encounter change. This dynamic between game and population can be formalized. In doing so, we explore a pair of questions: how do the actions of individuals depend on the local social context and, in turn, how does the social context change as a direct result of strategic actions taken by individuals? These questions are at the core of understanding the short- and long-term coupling of behavior and environment.

To begin, consider the strategy-dependent payoff of the hawk-dove game competing for nesting sites, which can be written as follows:

		Player 2: Hawk	Player 2: Dove
Player 1	Hawk	$\left(\frac{G-C}{2}, \frac{G-C}{2}\right)$	$(G, 0)$
	Dove	$(0, G)$	$\left(\frac{G}{2}, \frac{G}{2}\right)$

G denotes the gain associated with a nesting site, and C denotes the cost of fighting. The payoffs are symmetric, such that the first number in each payoff pair is the payoff received by the player in the row and the second number in each payoff pair is the payoff received by the player in the column. For example, when hawks play doves, then doves flee and the hawks receive the entire benefit, G, whereas doves suffer no cost and receive a payoff of 0. In contrast, when two hawks encounter a nesting site, then they fight and, on average, each

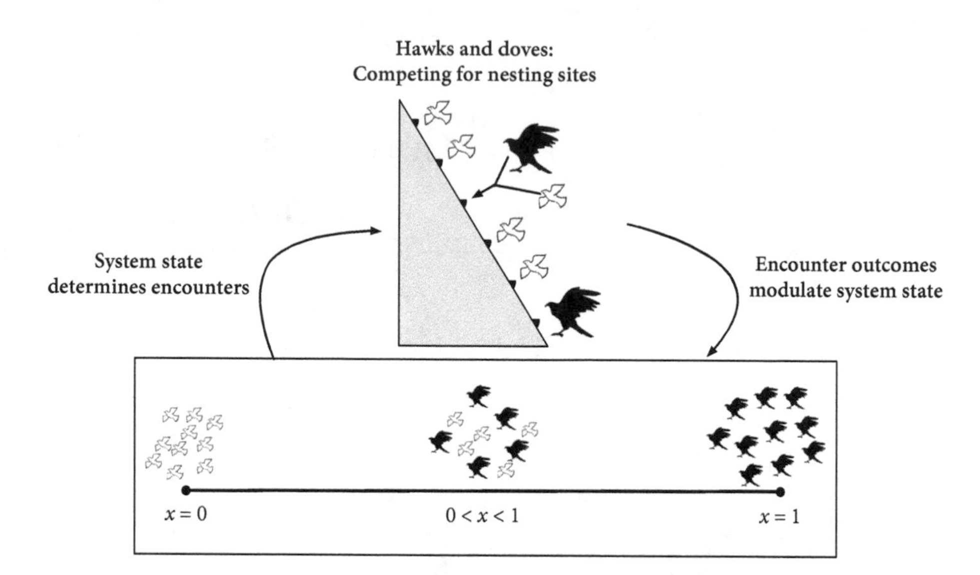

Figure 10.1: Hawk-dove games in which the population state (as determined by the fraction of hawks and doves) sets the context for games. Here games denote interactions between a pair of individuals for a nesting site. Then the outcome of the game changes the relative fraction of hawks and doves.

hawk wins the site (with value G) and loses the fight (at a cost C) one-half of the time. In the event that fights are costly, such that $C > G$, then it is possible for hawks and doves to coexist.

The standard replicator dynamics for two-player games can be written as

$$\dot{x}_1 = r_1(x, \mathbf{A})x_1 - \langle r\rangle(x, \mathbf{A})x_1 \tag{10.1}$$

$$\dot{x}_2 = r_2(x, \mathbf{A})x_2 - \langle r\rangle(x, \mathbf{A})x_2 \tag{10.2}$$

where r_1, r_2, and $\langle r\rangle$ denote the fitness of player 1, the fitness of player 2, and the average fitness, respectively, all of which depend on the frequency of players and the game-theoretic payoffs. The frequencies of strategies are denoted by x_1 and x_2. For example, in the case of hawks and doves, one can think of x_1 as the frequency of hawks and x_2 as the frequency of doves. Because $x_1 + x_2 = 1$, then we only need one variable, i.e., $x \equiv x_1$, to fully describe the state of the population. How then do we calculate fitness? In replicator dynamics, the fitness is directly linked to payoffs (for more details, see Hofbauer and Sigmund 1998; Nowak 2006).

For the hawk-dove game, hawks receive a payoff of

$$P_{\text{hawk}} = \overbrace{x \times (G - C)/2}^{\text{hawk–hawk}} + \overbrace{(1 - x) \times G}^{\text{hawk–dove}}, \tag{10.3}$$

which is equivalent to

$$P_{\text{hawk}} = G - \left(\frac{G + C}{2}\right)x. \tag{10.4}$$

This form reflects the fact that payoffs to hawks decrease as hawk frequency increases, precisely because the increase in hawk-hawk encounters leads to a higher frequency of (costly)

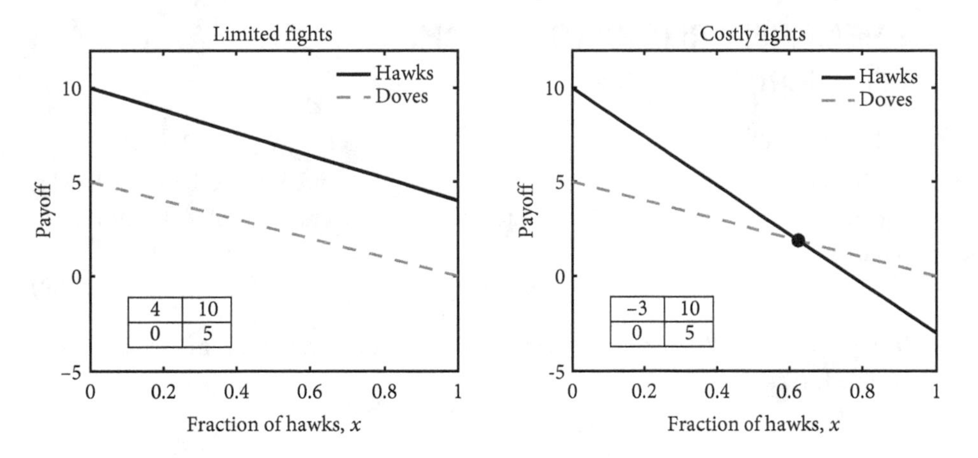

Figure 10.2: Both panels depict how payoffs in the hawk-dove game depend on social context, measured in terms of the fraction of hawks, x. (Left) Limited fights, where $G = 10$ and $C = 2$. (Right) Costly fights, where $G = 10$ and $C = 16$.

fights. In contrast, doves receive a payoff of

$$P_{\text{dove}} = \overbrace{x \times 0}^{\text{dove–hawk}} + \overbrace{(1-x) \times G/2}^{\text{dove–dove}}, \tag{10.5}$$

which is equivalent to

$$P_{\text{dove}} = \frac{G}{2}(1-x). \tag{10.6}$$

This form reflects the fact that doves interact with hawks a fraction x of the time but receive no payoff from such interactions, and interact with doves a fraction $1-x$ of the time and receive a payoff of $G/2$. As is apparent, both payoffs are context dependent, i.e., they depend on x. In replicator dynamics, the payoffs define a *frequency-dependent fitness*, i.e.,

$$r_1 \equiv P_{\text{hawk}} \tag{10.7}$$

$$r_2 \equiv P_{\text{dove}} \tag{10.8}$$

This regime can be seen in Figure 10.2.

The payoffs to hawks and doves raise questions at the center of this computational laboratory. First, how does the relative value of payoff curves influence the long-term dynamics arising from interactions between players, e.g., as in the hawk-dove game? Second, even if a certain outcome seems inevitable in the idealized "replicator dynamics" framework, is that always the case where outcomes are dictated, in part, by stochastic effects? Finally, when individuals interact in space, then there is the very real possibility that individuals are not chosen at random, but rather are highly correlated in space and in time. The lab addresses this last problem by moving away from hawks and doves and providing the guidelines to build a model directly inspired by recent work on interstrain killing by bacteria of the species *Vibrio cholerae*. As will be apparent, interstrain killing gives rise to the emergence of local patches, which, paradoxically can lead to the enhancement of cooperation.

10.2 MEAN FIELD REPLICATOR DYNAMICS OF MICROBIAL GAMES

10.2.1 Limited fights

Consider the dynamics of the hawk-dove (HD) game given interactions between hawks, with frequency $x \equiv x_1$, and doves, with frequency $1 - x \equiv x_2$. The replicator dynamics can be written as

$$\dot{x} = r_1(x, \mathbf{A}) - \langle r \rangle (x, \mathbf{A}) x \tag{10.9}$$

where $\langle r \rangle = r_1 x + r_2(1 - x)$ denotes the average fitness. To implement this model, develop a function called `hd_model` in Python.

```
def hd_model(x,t,pars):
    # function dxdt = hd_model(t,x,pars)
    #
    # Hawk-dove replicator dynamics model
    # Returns dxdt for a 1D replicator model with hawk-dove parameters,
    # pars.G and pars.C denoting gain and cost, in the matrix pars.A
    r1=x*pars['A'][0,0]+(1-x)*pars['A'][0,1]
    r2=x*pars['A'][1,0]+(1-x)*pars['A'][1,1]
    r_avg=r1*x + r2*(1-x)
    dxdt= r1*x - r_avg*x
    return dxdt
```

Applying this model requires choosing a payoff matrix. The following challenge problem will help you invoke this function to simulate hawk-dove games.

CHALLENGE PROBLEM: Hawk-Dove Simulation

Set up the main code to initialize the dynamics and loop through time for $G = 10$ and $C = 2$. A skeleton for the code is below, with missing pieces noted as . . . for you to fill in. If you want an extra challenge, try to superimpose the results of multiple simulations for the same gain, $G = 10$, but with three different costs: $C = 2$, $C = 6$, and $C = 9$ (as seen in the figure below).

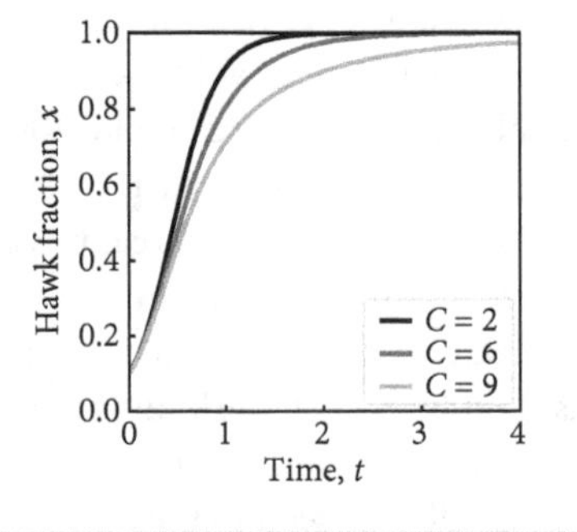

```
# Main simulation code
from _ _future_ _ import division
import matplotlib.pyplot as plt
import numpy as np
from scipy.integrate import odeint

G=10
C=2
A=np.array([[... , ...], [... , ...]])    # Fill in here
x0=0.1  # Initial fraction of hawks
tmax=4
t= np.linspace(0,tmax,10*tmax)

x =odeint(hd_model, x0, t, args=(A,G)) # Plot the results

plt.plot(t,x,label=r'$C={cost}$'.format(cost=pars['C']))
plt.xlabel('Time, t')
plt.ylabel('Hawk fraction, x')
plt.legend()
```

As is apparent, irrespective of the value of C, as long as $C < G$, then the dynamics converge to a system in which hawks predominate. It should also be apparent that as $C \to G$ it takes longer for hawks to eliminate doves. This observation forms the basis for the next section.

10.2.2 Costly fights and convergence

What happens when fights are costly, i.e., $C > G$? In that case, in a world of hawks, a hawk-hawk encounter has a net negative payoff of $-(C-G)/2$. In contrast, when a dove encounters a hawk, the dove leaves, with a payoff of 0. Sometimes not fighting is the best choice of all! In theory, the system should converge to a coexistence steady state including both hawks and doves, i.e., $0 < x^* < 1$. The fixed point must satisfy the condition $\dot{x} = 0$, i.e., when $r_1(x, A) = \langle r \rangle(x, A)$—when the fitness of hawks is equal to the average fitness. Another way to think about this is to recognize that the equilibrium occurs when the payoff to a hawk is equal to that of a dove. That occurs when $P_{dove}(x) = P_{hawk}(x)$, or when

$$\frac{G}{2}(1-x) = G - \left(\frac{G+C}{2}\right)x \tag{10.10}$$

such that

$$x^* = \frac{G}{C}. \tag{10.11}$$

This observation also forms the basis for the next challenge problem.

CHALLENGE PROBLEM: Hawk-Dove Dynamics

Using the values of $G = 10$ and $C = 16$, calculate the expected equilibrium; then simulate the hawk-dove game from multiple different initial conditions and demonstrate that all dynamics converge to the same fixed point. (Hint: Use the code from the previous challenge problem, but this time modify the loop to change the initial conditions rather than the game-theoretic payoffs.)

10.3 STOCHASTIC VERSIONS OF MICROBIAL GAMES

The prior section explained how to develop and analyze a replicator dynamics model of the hawk-dove game. The dynamics unfold assuming that the mean field limit applies. This term—mean field—describes the averages over many potential realizations of the game. But, in any particular instance, the stochastic changes of individual birth and death events can lead to very different population-level outcomes. This section reviews some of the basic theory of how to convert a continuous model into a discrete event simulation in which you can compare and contrast ensembles of stochastic simulations to replicator dynamics.

10.3.1 Some theory (a reminder from earlier in the guide)

As a refresher from earlier in the book: consider an event that occurs at a rate r, in units of inverse time, e.g., hrs^{-1}. There tends to be significant misunderstanding regarding the meaning of this kind of inverse unit. The inverse of the rate is a characteristic time, approximately equal to that of the time between random events. So events with a rate of 4 hrs^{-1} occur about every 15 minutes, whereas events with a rate of 10 hrs^{-1} occur about every 6 minutes. *Higher rates imply short intervals between events.* Formally, this rate should be thought of as a *probability per unit time* that the event occurs. The probability that an event occurs in some small interval dt is $r \times dt$. The probability that an event does not occur is $1 - r \times dt$. Hence, as long as we select the time interval to be sufficiently small, then $r \times dt < 1$ is the probability of the event taking place in the unit of time dt. If we interpret the fitness values as probabilities per unit time, then we can interpret positive values as the probability of birth and negative values as the probability of death.

10.3.2 Events and frequency dependence—stochastic hawk-dove games

Consider an environment in which there are a finite and fixed number of players, N. These individuals can be either hawks or doves. We define a game step as the result of N random games each played between a random pair of individuals. The focal player is the first player chosen and the opponent is the second player chosen. There can be four kinds of games: hawk vs. hawk, hawk vs. dove, dove vs. hawk, and dove vs. dove, only three of which are categorically distinct. The payoffs of these four games can be read directly off the payoff matrix; e.g., in the case of $G = 10$ and $C = 16$, the case of costly fights, the payoff matrix is as follows:

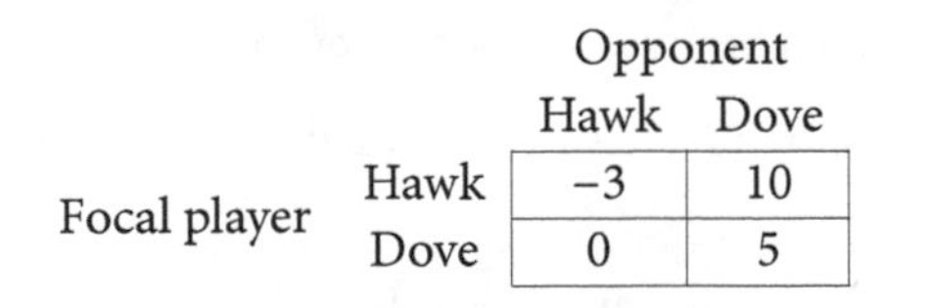

		Opponent	
		Hawk	Dove
Focal player	Hawk	−3	10
	Dove	0	5

If the number of individuals is finite, how do we change the state of the system after a game is played? This question forms the basis for going from theory (and concept) to an implementation that hews closer to biological reality.

Consistent with the replicator treatment, we interpret the payoff values as probability rates per unit time of either birth (when payoffs are positive) or death (when payoffs are negative). But birth would lead to an increase in the number of individuals and death would lead to a decrease in the number of individuals. Yet the replicator dynamics only deal in *frequencies*. Hence, to be consistent, we consider a stochastic hawk-dove game where the total number of players is fixed. To do so, whenever there is a birth, it must take over the spot of a current player (at random). Similarly, whenever there is a death, one of the remaining players must take over the vacated spot of the removed individual. The theoretical basis for why this sequence of moves recapitulates replicator dynamics is described in the main text. But the proof is in the pudding. The following three graphs illustrate this concept in action, using different-size populations. Two points are striking. First, the stochastic model really does recapitulate the mean field dynamics (visually at least; you can try many runs and take an average if you want to convince yourself further). Second, the size of the population matters, in terms of setting the magnitude of fluctuations (small populations have large fluctuations and vice versa).

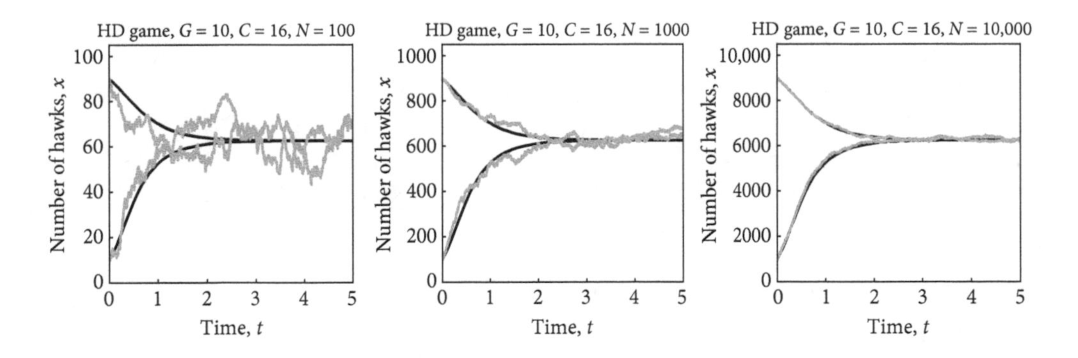

Yet this sequence of images is nontrivial. And why would you trust us? You should convince yourself, by building your own stochastic game, itself a gateway to the spatial type VI secretion model that serves as the capstone of this laboratory.

10.3.3 Pseudocode for stochastic HD games

As before, we use "pseudocode" to build up the structure of a simulation and then fill in the parts with actual code. This is particularly useful when the code itself has modules that can be tested independently of the whole. This is a useful way to think about simulations and also about problems as a whole both for prethinking and for bug testing after the fact. Here is a pseudocode for the simulation:

Loop for each time At each time step do the following:

Play N games Each game step includes N games, in which the following occurs:

- Choose a pair of players at random, call them p_1 and p_2, for focal player and opponent (or player 1 and player 2).
- Given the identities of players 1 and 2, find the payoff for the focal player only; call this a.
- If the payoff is positive, $a > 0$, then the focal player reproduces with probability $a \times dt$ and replaces a randomly chosen individual with an offspring.
- If the payoff is negative, $a < 0$, then the focal player dies with probability $|a| \times dt$ and is replaced by an offspring of a randomly chosen individual.
- Critically, offspring of any individual has the same type, i.e., hawks beget hawks and doves beget doves.

Store the number of hawks After each game step, record the number of hawks, N_h, and the fraction of hawks, $x = N_h/N$.

Converting this pseudocode into actual code takes some time (and experience). Hence, the next section gets you started, providing a solid basis for comparison with the replicator dynamics.

10.3.4 Stochastic HD games—modular implementation

Assuming that you still have the same structure as used before, consider the following new fields:

```
pars={}
pars['N']=1000
pars['G']=10
pars['C']=16
pars['A']=np.array([
        [(pars['G']-pars['C'])/2, pars['G']],
        [0, pars['G']/2]])
pars['x0vals'] = [0.1, 0.9]
pars['num_iter']=500
pars['dt']=0.01
pars['tmax']=pars['dt']*pars['num_iter']
```

With these in place, here is how you might want to build out the components.

Choosing random players In order to choose one individual, one can think of the identities of each individual, or simply the fraction of current players that are hawks. The latter implementation can lead to a few questions, e.g., can a player interact with itself? Assuming it can simplifies the game play, but if you want to keep track of the identity of each player, that is okay. First, it is important to establish a few conventions: the identity 0 refers to hawks and the identity 1 refers to doves—this also coincides with the row/column notation of the payoff matrix. Hence, if x is the current fraction represented by the variable `curx`, then consider the following code snippet:

```
p1=np.random.uniform()<curx
tmpid1=1-(p1==1)
p2=np.random.uniform()<curx
tmpid2=1-(p2==1)
```

This code snippet takes a random number between 0 and 1 and compares it to the value of `curx`. If it's smaller, then the focal player is a hawk. Hence, the next command maps a positive result from the random number choice to the iD 0 (denoting hawks) and a negative result to the iD 1 (denoting doves). Replicated twice, the code now has the identities of the focal and opponent players. These identities also form the basis for finding the payoff for this game.

Find the payoff Given the above, the code should have variable terms `tmpid1` and `tmpid2` for the hawk/dove identity of the focal and opponent players. Hence, the payoff should just be

```
payoff = pars['A'][tmpid1,tmpid2]
```

With this step, the question is how to deal with the positive and negative cases.

Birth events in the stochastic HD game There are two cases to consider: either the focal player is a hawk or it is a dove. In either case, the payoff could be positive or negative (of note, we ignore the 0 case, assuming that nothing happens). Let's focus on the hawk case first in the event that the payoff is positive, which will happen when hawks interact with doves or when hawks interact with hawks but the fights are not costly. The following code snippet represents the treatment of this case—again assuming that `curx` is defined. Note that there is only one scenario when the game leads to a change in frequencies, i.e., when a hawk gives birth and replaces a dove, in which case the hawk frequency goes up by a value of $1/N$ (because there is now one more hawk out of N in the population).

```
if payoff>0:
    birth = np.random.uniform() < payoff*pars['dt']
    if birth:
        replace_dove = np.random.uniform() >= curx
        if replace_dove:
            curx = curx+1/pars['N']
```

This code snippet forms the basis for the entire code base, which you can develop next.

10.3.5 Stochastic HD games—the full challenge

The full stochastic hawk-dove game is built on the components in the previous section. Here then is your challenge: to build a stochastic hawk-dove game. There are many ways to do this; hence, the following code snippet is modular. But it is up to you to decide whether you want to use this structure or come up with your own. With the full model in hand, there are many kinds of problems to consider:

- How does the number of individuals affect the magnitude of fluctuations at equilibrium?
- In the stochastic case, will hawks always win if $C < G$? To answer this, consider starting with $N = 1000$ and $x0 = 5/1000$, or 0.005.
- Does the average of many stochastic simulations recapitulate the replicator dynamics?

As just one example, below are two simulations of games that should end with coexistence (on the left) and hawk domination (on the right), but in a stochastic model sometimes the system leads to the complete elimination of hawks. This is just one example of differences that emerge as you move from mean field theory to individual- and stochastic-based games.

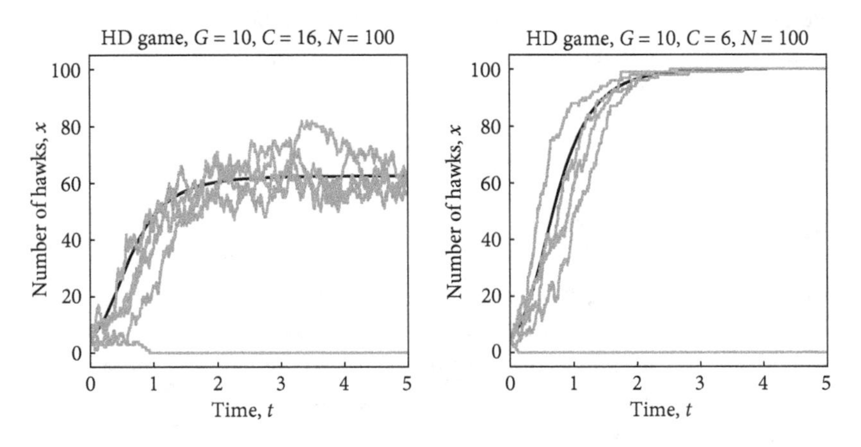

The code block in the following problem represents the challenge. The solution follows (but try to figure out as much as you can on your own!).

CHALLENGE PROBLEM: Stochastic Hawk-Dove Game

Fill in the following components and build your own stochastic hawk-dove game.

```
# Initialize
stats={}
curx = pars['x0']
x=[pars['x0']]
t=[0]
# Run for each time
for i in range(pars['num_iter']):
    # Run for each player
    for j in range(pars['N']):
        p1 = np.random.uniform()<curx # Identity of player 1
        tmpid1 = 1-(p1==1) # In payoff matrix: hawk is 0, dove is 1
        p2 = np.random.uniform()<curx # Identity of player 2
        tmpid2 = 1-(p2==1) # In payoff matrix: hawk is 0, dove is 1
```

```
        # Game cases
        if p1==1: # Hawk
            payoff = pars['A'][tmpid1,tmpid2]
            if payoff>0: # Hawk may give birth
                birth = np.random.uniform() < payoff*pars['dt']
                if birth:
                    replace_dove = np.random.uniform() >= curx
                    if replace_dove:
                        curx = curx+1/pars['N']
            elif payoff<0: # Hawk may die
                death = ... Fill in ...
                if death:
                    ... Fill in ...
        else: # p1 is dove
            ... Fill in ...
    # After looping through all N players
    x.append(curx)
    t.append(t[-1]+pars['dt'])
stats['x']=np.array(x)
stats['t']=np.array(t)
```

10.4 TYPE VI SECRETION—A KILLER GAME, IN SPACE

10.4.1 A minireview of the system

The basis of this system is discussed in detail in the textbook. As a reminder, *Vibrio* strains use a type VI secretion to transport lethal proteins via a spike/syringe system to adjacent cells. Cells of the same strain express an antitoxin that renders the toxin ineffectual (though it comes at a cost to growth). These toxins can kill adjacent cells that are unprotected, e.g., because they do not express an antitoxin or express an incompatible antitoxin. As shown by McNally et al. (2017), this simple rule set is sufficient to give rise to nontrivial spatial patterns, including the "coarsening" of a system from random to coherent patterns.

10.4.2 A game-theoretic version

Consider the strategy-dependent payoff of two *Vibrio* strains that compete in a two-dimensional spatial habitat, e.g., an agar plate, which can be written as follows:

		Player 2: Red	Player 2: Blue
Player 1	Red	$G/2$	$(G-C)/2$
	Blue	$(G-C)/2$	$G/2$

Here G denotes the gain associated with reproducing, and C denotes the cost of the type VI killing machinery by a noncompatible type. The use of the terms "red" and "blue" is consistent with the choice in the collaborative study of McNally et al. Type VI secretion systems (T6SS) lead to the potential death of the target cell. Although it would seem that costly fights are essential to this problem, any level of cost is sufficient such that $C > G$.

Via the replicator dynamics notation for two-player games, we can write the expected dynamics of the red type with frequency x as

$$\dot{x} = r_1(x, \mathbf{A}) - \langle r \rangle (x, \mathbf{A})x \tag{10.12}$$

and the expected dynamics of the blue type with frequency $1 - x$ as

$$r_1(x) = Gx/2 + (1-x)(G-C)/2 \tag{10.13}$$

$$r_2(x) = (G-C)x/2 + (1-x)G/2 \tag{10.14}$$

$$\langle r \rangle = r_1 x + r_2(1-x) \tag{10.15}$$

After some algebra, this can be rewritten as

$$\dot{x} = x(1-x)(r_1(x) - r_2(x)) \tag{10.16}$$

or, after some reduction (as shown in the main text),

$$\dot{x} = Cx(1-x)\left(x - \frac{1}{2}\right). \tag{10.17}$$

This cubic equation has three roots, at 0, 1/2, and 1. As such, we expect that for any initial conditions where $x < 1/2$ the system will be driven into an "all-blue" state, whereas for any initial conditions where $x > 1/2$ he system will be driven into an "all-red" state.

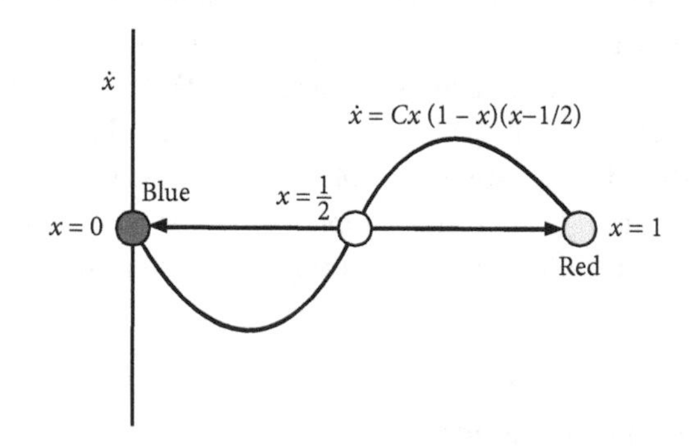

This notion of alternative stable states underlies the spatial, stochastic game that we explore next.

10.4.3 Spatial T6SS Game

You are now ready to build your own microbial spatial game. Impossible? Absolutely not. In fact, this model will be simpler in many ways than the hawk-dove game, particularly given that many of the same concepts will be used. Plus, the payoff in an altogether different sense

Figure 10.3: Spatiotemporal dynamics of a T6SS game in space, from random initial conditions to labyrinthine patterns. This game uses $G = 10$ and $C = 18$. The times shown here are after 0.2 time units (the equivalent of 20 iterations, each with $dt = 0.01$).

is quite rewarding, in that when the model works, you will be able to see still images (and videos) that look very much like the images in Figure 10.3.

Now, how to do it. First, the pseudocode concept helps. But before getting there, it is evident we need a slightly different framework, i.e., one in space.

10.4.4 Pseudocode for spatial, stochastic HD games

Here is a pseudocode for the spatially explicit simulation:

Loop for each time At each time step do the following:

Play *N* games Each game step includes *N* games, in which the following occurs:

- Choose a focal player—player 1.
- Choose an opponent player at an adjacent site—player 2.
- Given the identities of players 1 and 2, find the payoff for the focal player only; call this a.
- If the payoff is positive, $a > 0$, then the focal player reproduces with probability $a \times dt$ and replaces a randomly chosen adjacent individual with an offspring.
- If the payoff is negative, $a < 0$, then the focal player dies with probability $|a| \times dt$ and is replaced by an offspring of a randomly chosen, adjacent individual.

- Critically, offspring of any individual has the same type, i.e., red begets red and blue begets blue.

Note that perhaps the most critical feature difference here is that player 2 is selected from an *adjacent* site to player 1. The fact that players only interact locally forms the basis for the difference between this model and the prior model.

10.4.5 Spatial T6SS games—components

There are a few components to make this game work. The pieces are as follows:

Initialize a random habitat A random habitat can be implemented by generating a two-dimensional array with random values, 0 for red and 1 for blue. The following code snippet will rapidly generate environments with a 50-50 split in types:

```
L = 200
x1 = 0.5
x=np.random.choice([0,1],[L,L],[0.5,0.5])
```

Visualize the habitat, sensibly Once you have a habitat, it would be nice to make the red types red and the blue types blue. To do so requires understanding something about colormaps. In brief, colormaps are a way to translate an array value into a pixel value using a three-channel RGB (red-green-blue) color scheme. In our case, we keep things simple and use the `image` command, which takes the integer values in an array literally. As part of its built-in visualization scheme, the interpreter will then look for a set of three numbers in a `colormap` array as the RGB pixel values. For example, try this:

```
plt.imshow(x,cmap='RdYlBu')  # Visualize the habitat
```

Find the adjacent player Consider the following definition of an array: `moves = np.array([[-1,0],[1,0],[0,-1], [0,1]])`, which when visualized looks as follows:

```
In[160]: moves
Out[160]:
array([[-1,  0],
       [ 1,  0],
       [ 0, -1],
       [ 0,  1]])
```

These are the relative positions of the adjacent player, either one to the left, one to the right, one below, or one above of the current position. Hence, one way to randomly select an adjacent player is to do the following:

```
foc_pos = np.random.choice(range(L),[1,2])[0]
opp_pos = foc_pos + moves[np.random.choice(range(4))]
opp_pos = pbc_reset(opp_pos,L)    # Accounts for PBCs
```

This code takes an (x, y) coordinate of the focal player and adds a randomly selected shift to it to identify the adjacent player. All seems well, with one exception: boundary conditions. What choice of "neighbor" should one use when the focal player is already at the boundary? For various reasons, we recommend the use of periodic boundary conditions, implicit in the function `pbc_reset` shown here.

Implement periodic boundary conditions Periodic boundary conditions imply, that although the actual habitat may be a flat 2D space, it is often preferable to pretend as if the dynamics take place on a torus (think of the surface of a doughnut). This tends to remove finite size effects because every point is just as much in the middle as any other. There are many solutions to this problem, but for now, the next challenge is to write your own code to shift the positions, assuming a square lattice of size L in each dimension.

CHALLENGE PROBLEM: Periodic Boundary Conditions

Devise a solution to the problem of taking in a 2D position and returning a position that "wraps around" in a torus-like sense. You should write a function that accepts a 2D position and the size of the grid and returns a new position.

10.4.6 Spatial T6SS games—ready, player 1?

It's time for the final challenge. Are you ready, player 1? Here goes.

CHALLENGE PROBLEM: T6SS Stochastic Spatial Model

Write your own T6SS stochastic spatial dynamics. If you want this challenge without any further help, then stop reading. In essence: spoiler alert ahead! But, if you would like some assistance, then see the solution that follows and copy/paste as little or as much as you'd like. Recall that you will need to set up the colormap as noted above.

```
from __future__ import division
import matplotlib.pyplot as plt
import numpy as np

L=200
N=L**2
A=np.array([[5,-4],[-4,5]])
x1=0.5
```

```
num_iter=500
dt=0.01
tmax=dt*num_iter
moves = np.array([[-1,0],[1,0],[0,-1], [0,1]])
x=np.random.choice([0,1],[L,L],[0.5,0.5])

for i in range(num_iter):
    for j in range(N):
        foc_pos = np.random.choice(range(L),[1,2])[0]
        foc_id=...
        opp_pos = foc_pos + moves[np.random.choice(range(4))]
        opp_pos = pbc_reset(opp_pos,L)
        opp_id=...
        payoff=...
        if payoff > 0:
            birth = np.random.uniform() < payoff*dt
            if birth:
                rep_pos=foc_pos+moves[np.random.choice(range(4))]
                rep_pos=pbc_reset(rep_pos,L)
                x[rep_pos[0],rep_pos[1]]=...
        elif payoff < 0:
            death = np.random.uniform()<np.abs(payoff)*dt
            if death:
                rep_pos=foc_pos+moves[np.random.choice(range(4))]
                rep_pos=pbc_reset(rep_pos,L)
                x[foc_pos[0],foc_pos[1]]=...
    if np.mod(i,40)==0 or i==199:
        plt.figure()
        plt.axis('off')
        plt.title('Gen %d'%i)
        plt.imshow(x,cmap='RdYlBu')
        plt.savefig('Gen %d'%i, dpi=300)
```

As should be apparent, this laboratory is (slightly) more extensive than prior labs. However, the concepts should be relatively familiar at this stage. Like earlier laboratories, it is possible to consider individual interactions as the basis for coupled systems of nonlinear ODEs; i.e., these can be understood as the expected value of populations in a stochastic model. However, the deeper point reinforced by this laboratory is that it is possible to use relatively simple coding strategies and represent models that incorporate stochasticity arising from individual (vs. continuous) dynamics as well as from explicitly spatial interactions. Notably, the generalization of the model in this laboratory opens the door to many microbial and nonmicrobial games alike.

SOLUTIONS TO CHALLENGE PROBLEMS

SOLUTION: Hawk-Dove Simulation

The key code to fill in is as follows, recognizing that the order of rows and columns is critical to making sure that the payoffs are accurately reflected in the matrix:

```
import matplotlib.pyplot as plt
import numpy as np
from scipy.integrate import odeint

CList=[2,6,9]
for i in range(3):
    pars['C']=CList[i]
    pars['G']=10
    pars['A']=np.array([[(pars['G']-pars['C'])/2, pars['G']], [0, pars['G']/2]])
    x0=0.1
    pars['tmax']=4;
    t=np.linspace(0,pars['tmax'])
    # Run the model
    x =odeint(hd_model, x0, t, args=(pars,))
    plt.plot(t,x,label=r'$C={cost}$'.format(cost=pars['C']),
             color=i*np.array([0.3,0.3,0.3]),linewidth=3)    # Plot results
plt.xlabel(r'Time, $t$',fontsize=20)
plt.ylabel(r'Hawk fraction, $x$',fontsize=20)
plt.xlim([0,4])
plt.ylim([0,1])
plt.legend(loc = 'best')
```

SOLUTION: Hawk-Dove Dynamics

The expected fixed point is $x^* = 10/16 = 5/8$, which is equal to 0.625. The code snippet is found below, with a small addition, overlaying a single circle at the expected equilibrium.

```
from _ _future_ _ import division
import matplotlib.pyplot as plt
import numpy as np
from scipy.integrate import odeint

# Hawk-dove dynamics code
G=10
C=16
tmax=4
A=np.array([[(G-C)/2,G],[0,G/2]])
```

```
t= np.linspace(0,tmax,50)

# Simulate the game and plot

for x0 in [0.1, 0.3, 0.7, 0.9]:
    x =odeint(hd_model, x0, t, args=(A,G))
    plt.plot(t,x,label='$x_0$=%s'%x0)    # Plot results

plt.plot(tmax, G/C, 'ko')        # Plot the expected fixed point
plt.title('Costly Fights, G=10, C=16')
plt.xlabel('Time, $t$')
plt.ylabel('Hawk fraction, $x$')
plt.xlim([0,4.05])
plt.ylim([0,1])
plt.legend()
```

The resulting dynamics look like the plot on the left, with the equilibrium value as a function of C plotted on the right. Both graphs use the same gain value of $G = 10$.

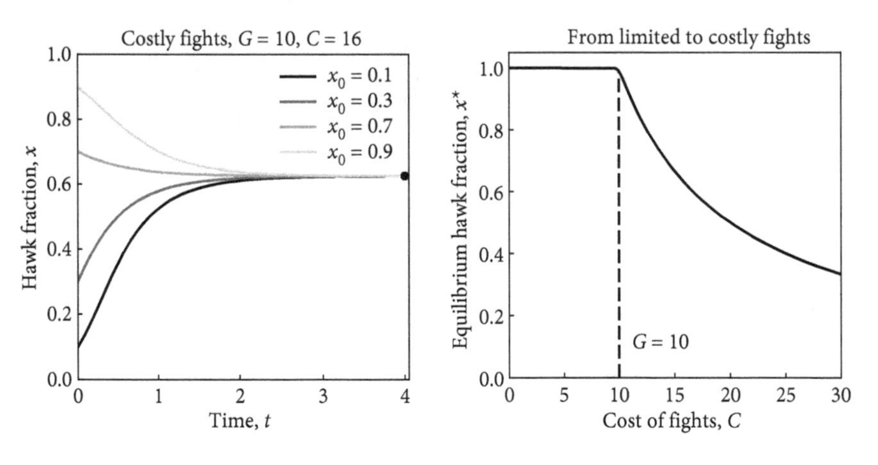

SOLUTION: Stochastic Hawk-Dove Game

The following includes components for each potential two-player case:

```
if p1==1: # Hawk
    payoff = pars['A'][tmpid1,tmpid2]
    if payoff>0:
        birth = np.random.uniform() < payoff*pars['dt']
        if birth:
            replace_dove = np.random.uniform() >= curx
            if replace_dove:
```

```
                    curx = curx+1/pars['N']
        elif payoff<0:
            death = np.random.uniform()< np.abs(payoff)*pars['dt']
            if death:
                replace_dove = np.random.uniform() >= curx
                if replace_dove:
                    curx = curx - 1/pars['N']
    else: # p1 is dove
        payoff = pars['A'][tmpid1,tmpid2]
        if payoff > 0:
            birth= np.random.uniform() < payoff*pars['dt']
            if birth:
                replace_hawk=np.random.uniform() < curx
                if replace_hawk:
                    curx=curx-1/pars['N']
        elif payoff < 0:
            death =np.random.uniform() < np.abs(payoff)*pars['dt']
            if death:
                replace_hawk=np.random.uniform()
                if replace_hawk:
                    curx=curx+1/pars['N']
```

SOLUTION: Periodic Boundary Conditions

One potential solution to the problem is as follows:

```
def pbc_reset(old_pos,L):
    # function new_pos = pbc_reset(old_pos,L)
    #
    # Assuming periodic boundary conditions shifts the
    # positions back into the grid 0..(L-1)
    new_pos=old_pos
    for i in range(2):
        if old_pos[i] >= L:
            new_pos[i]=0
    return new_pos
```

SOLUTION: T6SS Stochastic Spatial Model

Here is a working version of the T6SS stochastic spatial dynamics code. It can be modified to generate multiple stills, videos, and analysis of the properties of the spatiotemporal dynamics of types. How you modify it is up to you.

```
from _ _future_ _ import division
import matplotlib.pyplot as plt
import numpy as np

L=200
N=L**2
A=np.array([[5,-4],[-4,5]])
x1=0.5
num_iter=500
dt=0.01
tmax=dt*num_iter
moves = np.array([[-1,0],[1,0],[0,-1], [0,1]])
x=np.random.choice([0,1],[L,L],[0.5,0.5])

for i in range(num_iter):
    for j in range(N):
        foc_pos = np.random.choice(range(L),[1,2])[0]
        foc_id=x[foc_pos[0],foc_pos[1]]
        opp_pos = foc_pos + moves[np.random.choice(range(4))]
        opp_pos = pbc_reset(opp_pos,L)
        opp_id=x[opp_pos[0],opp_pos[1]]
        payoff=A[foc_id,opp_id]
        if payoff > 0:
            birth = np.random.uniform() < payoff*dt
            if birth:
                rep_pos=foc_pos+moves[np.random.choice(range(4))]
                rep_pos=pbc_reset(rep_pos,L)
                x[rep_pos[0],rep_pos[1]]=foc_id
        elif payoff < 0:
            death = np.random.uniform()<np.abs(payoff)*dt
            if death:
                rep_pos=foc_pos+moves[np.random.choice(range(4))]
                rep_pos=pbc_reset(rep_pos,L)
                x[foc_pos[0],foc_pos[1]]=x[rep_pos[0],rep_pos[1]]
    if np.mod(i,40)==0 or i==199:
        plt.figure()
        plt.axis('off')
        plt.title('Gen_{gen}'.format(gen=i))
        plt.imshow(x,cmap='RdYlBu')
        plt.savefig('Gen_{gen}.pdf'.format(gen=i), bbox_inches='tight')
```

Chapter Eleven

Eco-evolutionary Dynamics

11.1 FROM PREDATION EVENTS TO POPULATION DYNAMICS

The study of predator-prey dynamics lies at the very heart of quantitative biosciences. There are now over 100 years of history in the effort to understand how feedback between individual predator-prey events can lead to unexpected dynamics at the population scale. The core ideas are straightforward. Predators consume prey, thereby reducing prey densities and increasing predator densities. In turn, as predators become more abundant and prey less abundant, then predators have, relatively speaking, fewer prey to consume. As such, the predator populations begin to decline, reducing predation pressure and enabling the prey population to increase again. At this point with replete prey, the predators can thrive, and the cycle repeats. This cycle, reviewed in the main text, is known as the canonical predator-prey cycle, or the Lotka-Volterra cycle (Volterra 1926; Lotka 1925). It has as its hallmark a particular ordering of peaks and troughs; that is, first the prey peaks, then the predator peaks, then the prey reaches its lowest point, then the predator, and then the prey rebounds again. Yet, as helpful as 100 years of history can be, such canonical understanding can also be limiting. If we continue to assume that predator-prey cycles always behave a certain way, then we are less likely to see the data in the way that it should be interpreted, i.e., viewing key biological observations as "exceptions" rather than as mechanisms worthy of exploration.

What if predator peaks don't always follow prey peaks? What happens if, instead, such predator-prey dynamics are in phase, out of phase, or even "reversed" in orientation? Indeed, all of these "anomalous" features do occur in real systems, and not just seldom. The main text treats this issue at length, including the theoretical basis and evidence for non-canonical cycles in natural and experimental systems. The commonplace nature of such exceptions arises, in part, because two key assumptions of the canonical predator-prey dynamics model are rarely met (Figure 11.1). These violations of assumptions also provide a natural entry point for the laboratory.

The two key assumptions that we examine in this computational laboratory are the following:

The speed of evolution Many ecological models assume that populations are homogeneous, i.e., are made up of individuals that have the same phenotypes (at least with respect to predator-prey interactions). This in turn requires an implicit assumption

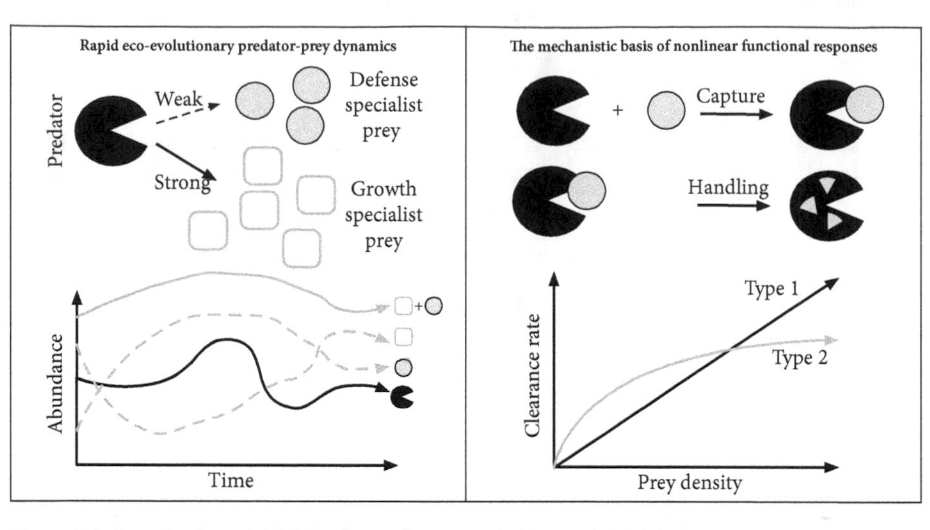

Figure 11.1: Organization of this laboratory, from population models including ecological and evolutionary dynamics (left) to the mechanistic basis of functional responses (right). As will be apparent, variation in the behavior of predators underlies both how many prey can be consumed and how the population and trait dynamics of predators and prey change.

about time scales, i.e., that evolutionary change is very slow compared to ecological dynamics. However, such assumptions are also violated, in that evolution can be rapid such that multiple genotypes of predators and/or prey can coexist at a given moment, and the total population dynamics of predator and prey can behave quite differently than expected.

The form of contact rates Population dynamics models often make an assumption that contact rates (and indeed clearance rates) of prey by predators is linear, when in fact such interaction rates are "nonlinear." These nonlinear interactions arise because of many biological mechanisms, including the time it takes to locate and handle prey.

There are of course many other directions we could choose to explore, from the effect of demographic stochasticity to space. In Section 11.2, the lab focuses on what can happen when more than one type of prey is present in a community, and how projecting a higher-dimensional system into a lower-dimensional space can help reconcile seemingly impossible violations of the canonical predator-prey dynamics. In Section 11.3, the lab explores the effects of functional responses on dynamics and then goes deeper, exploring the roots of functional responses via an individual-based model.

11.2 ECOLOGICAL DYNAMICS WHEN EVOLUTION IS FAST

11.2.1 Canonical predator-prey dynamics

The standard predator-prey dynamics model given N prey and P predators with prey density limitation is

$$\dot{N} = rN(1 - N/K) - bNP$$
$$\dot{P} = cNP - mP$$

Here r denotes the maximum prey growth rate, K is the prey carrying capacity, b is the clearance rate, c is the conversion rate from prey to predator (e.g., $c = \epsilon b$ where $\epsilon < 1$), and m is the predator decay rate in the absence of prey. This code can be separated into two parts: a model function and a script to evaluate dynamics. The model for this system can be written as:

```
def model_pp(y,t,pars):
   # Returns the rate of change of prey and predator in the standard LV model
   dydt = np.zeros(2)
   dydt[0] = pars['r']*y[0]*(1-y[0]/pars['K'])-pars['b']*y[0]*y[1]
   dydt[1] = pars['c']*y[0]*y[1]-pars['m']*y[1]
   return dydt
```

These dynamics can be shown to lead to a convergent equilibrium insofar as $N^* < K$ or equivalently $\frac{m}{c} < K$, such that predators can grow on levels of prey that are below the carrying capacity. The derivation of this result is found in the main text, and underlies the first challenge problem.

CHALLENGE PROBLEM: Convergent Predator-Prey Dynamics

Consider zooplankton (microscopic predators) consuming bacteria (their prey) such that $r = 1$ hrs^{-1}, $K = 10^7$/ml, $c = 0.1\,b$, and $m = 0.1$ hrs^{-1}. Find the critical value of the clearance rate b that enables predator-prey coexistence and show the results of dynamics as functions of time or in the phase plane (i.e., prey on the x axis and predators on the y axis).

11.2.2 Functional responses—a phenomenological view

The standard predator-prey dynamics model given N prey and P predators with prey density limitation and a generalized "functional response" is

$$\dot{N} = rN(1 - N/K) - bf(N)P$$
$$\dot{P} = cf(N)P - mP$$

where $f(N)$ denotes the functional response, i.e., how the clearance rate of prey by predators scales with prey density. The standard approach to classify such responses is shown below.

$$\text{type I}: f(N) \sim N \qquad (11.1)$$

$$\text{type II}: f(N) \sim \frac{N}{Q+N} \qquad (11.2)$$

$$\text{type III}: f(N) \sim \frac{N^2}{Q^2+N^2} \qquad (11.3)$$

These functional responses modify the stability of the coexistence fixed point (see the textbook for more details). As a means to explore this, consider a type II functional response in a particular predator-prey system:

$$\dot{N} = rN(1 - N/K) - \frac{bN}{1 + b\tau N}P \tag{11.4}$$

$$\dot{P} = \frac{\epsilon bN}{1 + b\tau N}P - mP \tag{11.5}$$

where τ is an effective handling time. This is known as the MacArthur-Rosenzweig model (Rosenzweig and MacArthur 1963). A modified code to implement this model is as follows:

```
def pp_dyn_type2(y,t,pars):
    # Predator-prey dynamics
    N=y[0]
    P=y[1]

    dydt = np.zeros(2)
    dydt[0] = pars['r']*N*(1-N/pars['K'])-pars['b']*N*P/(1+pars['b']*pars['tau']*N)
    dydt[1] = pars['eps']*pars['b']*N*P/(1+pars['b']*pars['tau']*N)-pars['m']*P

    return dydt
```

CHALLENGE PROBLEM: From Convergence to Limit Cycles

Consider zooplankton (microscopic predators) consuming bacteria (their prey) such that $r = 1$ hrs^{-1}, $K = 10^7$/ml, $b = 10^{-6}$ (ml-hr)$^{-1}$, $\epsilon = 0.2$, and $m = 0.1$ hrs^{-1}. Try to find a value of τ that yields limit cycle dynamics and then, if possible, modify τ to change dynamics so that they only exhibit convergent dynamics to a fixed point. If it works, your output should look like the following image, where the values of τ_1 and τ_2 are yours to discover!

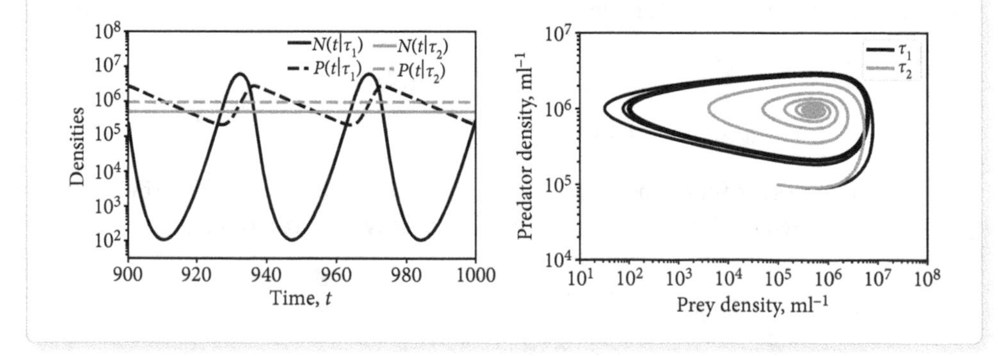

11.2.3 Steps toward rapid eco-evolutionary dynamics

The prior model assumed that interactions take place between a homogeneous population of prey and predators. However, what happens if there are, in fact, two distinct types of prey—one that is a growth specialist and one that is a defense specialist? It is often the case that such latent variation exists in the environment, even if it is not always possible to differentiate prey (or predators) in this way. The reasons are varied, including the possibility that genetic analysis is unfeasible, counts of types are based on morphological features (for which there is cryptic variation), and sites of variation may be unrelated to the predator-prey interaction. For example, if one were to use 16S rRNA metrics of diversity, then many bacteria would appear the same, even if their functional properties were radically different.

Hence, consider a variation of the standard predator-prey model in which there are two types of prey, N_1 and N_2, interacting with one predator:

$$\frac{dN_1}{dt} = r_1 N_1 \left(1 - \frac{N_1 + a_{12}N_2}{K_1}\right) - \frac{c_1 N_1 P}{(1 + \tau_1 c_1 N_1 + \tau_2 c_2 N_2)} \quad (11.6)$$

$$\frac{dN_2}{dt} = r_2 N_2 \left(1 - \frac{N_2 + a_{21}N_1}{K_2}\right) - \frac{c_2 N_2 P}{(1 + \tau_1 c_1 N_1 + \tau_2 c_2 N_2)} \quad (11.7)$$

$$\frac{dP}{dt} = \left(\frac{\epsilon c_1 N_1 + \epsilon c_2 N_2}{1 + \tau_1 c_1 N_1 + \tau_2 c_2 N_2}\right) P - mP \quad (11.8)$$

This is a natural extension of the single model; however, there is an important caveat. We may be interested in the case of dynamics for total prey and total predators, i.e., in the $N - P$ plane where $N \equiv N_1 + N_2$. Hence, the next challenge problem implements a variant of this model. If working, it should reveal *anti-phase* dynamics, where predators seem to peak at precisely the moment (more or less) when prey are reaching their trough. The reason for this is complicated and involves rapid switches in the relative frequency of prey genotypes, i.e., rapid evolution on ecological time scales! But some of that must be left for the homework. You may want to extend and modify this code to see the shift in genotypes for yourself.

CHALLENGE PROBLEM: Model of Eco-evolutionary Dynamics

Write the model described in Eq. (11.6) and run it with the following parameters:

```
# Parameters
pars={}
pars['r1']=1.8 # hr^-1
pars['r2']=1.75 # hr^-1
pars['c1']=3 # ml/hr
pars['c2']=0.3 # ml/hr
pars['tau'] = 1
pars['K1']=1.8 # Cells/ml
pars['K2']=1.75 # Cells/ml
pars['eps']=0.5 # Conversion
pars['m']=0.3 # hr^-1
```

The resulting figure should look like that on the next page; and note that a "stiff" integration method may help accelerate the simulations.

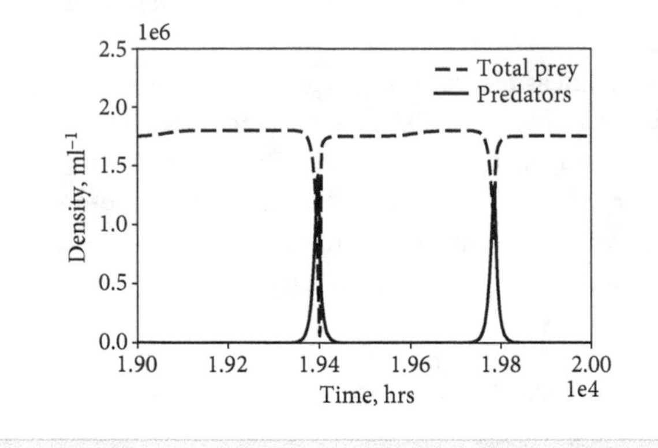

11.3 FUNCTIONAL RESPONSES—A MICROSCOPIC APPROACH

11.3.1 Theoretical background

Predators or consumers must interact with their prey or resource. Here we consider the "microscopic" basis for the kinds of functional forms used to describe predator-prey and consumer-resource models. Recall that the type I predator-prey model equations are

$$\dot{N} = rN(1 - N/K) - bNP$$
$$\dot{P} = cNP - mP$$

where the term bNP denotes consumption of prey by predators. This form suggests a collision-based approach, but does it also apply to spatial dynamics? Moreover, how would this kind of functional form change if predators had to "handle" hosts before consuming them? This lab and the homework will help you understand these questions.

First, let's assume that a predator moves "ballistically" through an environment with a velocity v. Moreover, assume the predator can sense prey in a zone of radius r. Hence, the predator will sense an area of size $\pi r^2 + 2rvT_s$ in a period T_s of searching (denoting the two separated caps of a sphere with a rectangular interior section). If prey are at a density N, then the potential number of prey encounters is $(\pi r^2 + 2rvT_s)N$. Further, if a fraction k are detected and of those f are consumed per unit time, then the total prey consumption is $\pi(r + vT_s)^2 fkT_sN$ per predator in a period of time T_s. This is true for all predators, as long as they tend *not* to run into or near each other. Hence, we assume that the total consumption per unit time scales as follows:

$$\text{consumption rate} = bNP$$

where the constant b approaches

$$\pi r^2 fk$$

in the limit that $T_s \to 0$. For example, you may find that this limit poses some problems, e.g., an immobile predator with relatively slow diffusing prey may rapidly deplete local prey. So even though it continues to sense across a radius of r, there may not be any prey left to sense! To test these ideas, we use an individual-based model (IBM) of a predator detecting and eating prey. This code should be downloaded first, given its complexity.

11.3.2 Getting started

First, download these files:

- `master_ibm.py`—the master script for setting parameters of the IBM model
- `ibm_functions.py`—dynamic simulator of predation as well as utility functions

Once you've downloaded the files, `master_ibm` in your notebook type so that the visualization displays in its own window. Then run the master script. You should see a predator with a zone of detection eating prey. The predator moves ballistically (in a straight line) and changes directions on occasion, whereas the prey move diffusively. Circles that appear denote the last set of prey eaten in a particular interval. These circles automatically update as new prey are consumed. Resize the window so you can see the update label. This is what it should look like when done:

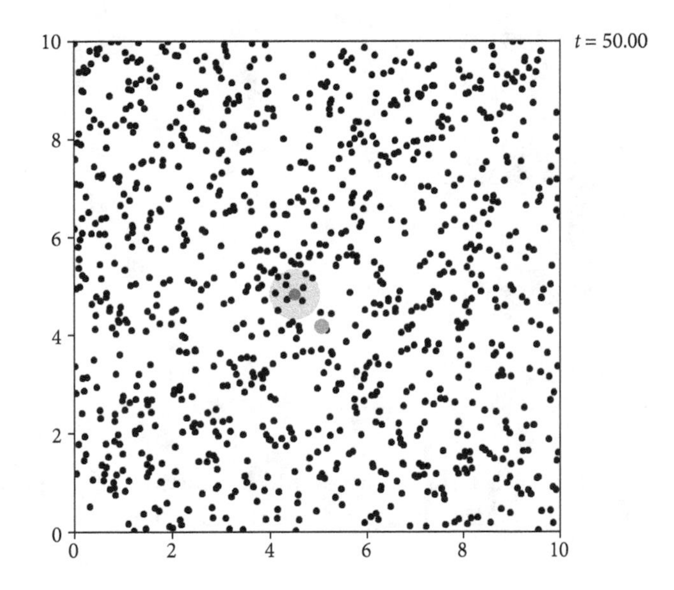

When the simulation is done, create a new figure window and type:

```
>> plt.plot(t,total_eaten)
```

which will give you a display of the cumulative number of prey eaten by the predator with time. Now that the code is working, let's take a closer look:

```
import numpy as np
import ibm_functions

# Basic information
info={}
info['prey_density']=10 # Density/cm^2
info['maxX']=10 # cm
```

```
info['maxY']=10 # cm
info['tf']=50 # sec
info['dt']=0.1 # Time step
info['replenish_prey']=1 # Should prey regenerate?
info['viz_dyn']=1 #1 for animation, 0 for no animation

# Define the prey
prey = {}
prey['num'] = info['prey_density']*info['maxX']*info['maxY']
prey['pos']=np.random.uniform(size=(prey['num'],2))
prey['pos'][:,0]=prey['pos'][:,0]*info['maxX']
prey['pos'][:,1]=prey['pos'][:,1]*info['maxY']
prey['diffusion']=0.005 # cm^2/sec

# Place the predator
predator={}
predator['pos'] = np.array( [[ info['maxX']/2, info['maxY']/2 ]] )
predator['theta']=np.random.uniform()*2*np.pi # Angle of movement
predator['r']=1.25 # Radius
predator['k']=0.3 # Detection
predator['f']=0.05 # Successful capture per time
predator['vel']=0.1 # cm/sec
predator['handling_time']=0.0 # sec
predator['tau']=6 # Run time length, sec
predator['trun'] = 0 # Initialize each run

# Simulate eating
reps=1
totend = np.zeros(shape=(1,reps))
for ens in np.arange(reps):
    print(ens)
    [t,numeaten] = ibm_functions.ibm_predation(info,predator,prey)
    total_eaten = np.cumsum(numeaten)
```

Do you understand what each term means? It will be worthwhile to spend some time looking over the somewhat difficult function `ibm_predation(info,predator, prey)`.

11.3.3 Exploration of predation and functional responses

The following are your objectives for the remainder of this section of the lab. Each is a challenge . . . with solutions revealed along the way.

CHALLENGE PROBLEM: Individual-Based Model Zone

Test the hypothesis that reductions in the zone of detection decrease the number of prey eaten by a squared factor. For example, set the predator radius equal to 1 and then, while keeping all other parameters the same, reduce the predator radius by half and see if there is a 4:1 ratio in the number of prey eaten.

CHALLENGE PROBLEM: IBM Diffusion

Modify the diffusion coefficient of prey. Does it modify the consumption rate of predators?

CHALLENGE PROBLEM: IBM Density

Change the density of prey (don't increase it beyond 50 or so). How does prey consumption change?

11.3.4 Lions and tigers and handling time, oh my!

Of course, predators do not instantaneously convert prey biomass into predator biomass (or even into new offspring). In the absence of an age-structured model with quotas, it is important to at least consider the effect of handling time. Here we need to restore all parameters to their original values and set

```
predator['handling_time']=0.5
```

This will make the predator wait 0.5 seconds for each prey it consumes. The model will continue to show which prey has been eaten (via a blue circle), even as the predator moves on. How does the total number of prey consumed compare to the prior case? Can you set parameters such that predators are essentially handling-time limited? (Hint: Set detection and fraction consumed to be nearly 1.) In which case does your model agree with predictions?

In the homework, you will explore how handling time changes the functional response. In particular, when predators must take time to handle and consume their prey, then the consumption rate is hypothesized to scale like this:

$$\text{consumption rate} = \frac{bNP}{1 + aN}$$

where b/a is equal to $1/T_h$ and T_h is the handling time. You will get a chance to use this individual-based model in the homework problems to explore if and how this hypothesis is valid.

SOLUTIONS TO CHALLENGE PROBLEMS

SOLUTION: Convergent Predator-Prey Dynamics

Predator-prey coexistence requires that $1 - N^*/K > 0$ or that $N^* < K$. Because prey densities are controlled by predators, $N^* = m/c$, which in this example is $N^* = 10\, m/b$, such that the condition on the clearance rate enabling predators to survive on prey is $b > \frac{10m}{K}$ or $b > 10^{-7}$ (ml-hr)$^{-1}$. The following figures show starkly different outcomes given choices of b, both 10-fold above and below that critical value.

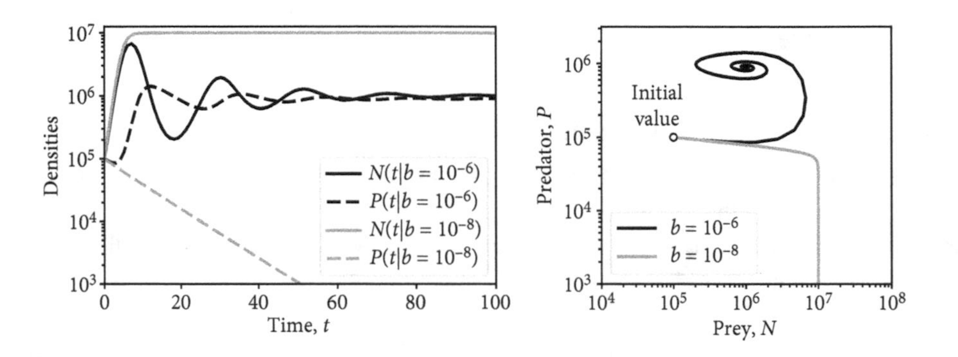

The phase plane figures are generated using the following code (legend optional):

```
# Units of hrs, ml
pars={}
pars['r']=1
pars['K']=10**7
pars['eps']=0.1
pars['m']=0.1
pars['T']=100

fig = plt.figure()
ax = fig.gca()

# Coexistence case
pars['b']=10**-6
pars['c'] = 0.1*pars['b']
t =np.linspace(0,pars['T'],100)
y0=10**5*np.ones(2)
y = integrate.odeint(model_pp,y0,t,args=(pars,))
plt.loglog(y[:,0],y[:,1],'k',linewidth=3)

# Failure case
pars['b']=10**-8
pars['c']=0.1*pars['b']
y = integrate.odeint(model_pp,y0,t,args=(pars,))
```

```
plt.loglog(y[:,0],y[:,1],color=[0.5,0.5,0.5],linewidth=3)
plt.loglog(y0[0],y0[1],marker='o',markeredgecolor='k',markerfacecolor='w')
plt.xlim(10**4,10**8)
plt.ylim(10**3,10**6.5)
plt.text(10**4.4,10**5.2,'Initial\nvalue',fontsize=16)
```

SOLUTION: From Convergence to Limit Cycles

The key to this challenge problem is to recognize that τ should have units similar to those of the growth rate. A detailed examination of the coexistence criteria and transition to limit cycles is in the main text. Here these examples are for $\tau_1 = 0.2$ (limit cycles) and $\tau_2 = 0.02$ (convergence to a fixed point). It is worth noting that, in the limit as $\tau \to 0$, the type II functional response converges to a type I functional response where limit cycles are not expected. The code for these simulations is similar to that above, modified given the invocation of different model functions. Hence, as an alternative, here is the code for the time series view, focusing only on the last 100 hours of the simulation:

```
t=np.arange(0,1000,0.1)
y0=10**5*np.ones(2)

fig=plt.figure()
ax = fig.gca()
# Limit cycle case
pars['tau']=0.2
y = integrate.odeint(pp_dyn_type2,y0,t,args=(pars,))
tmpi = np.where(t>900)
plt.plot(t[tmpi],np.transpose(y[tmpi,0]),'k',linewidth=3)
plt.plot(t[tmpi],np.transpose(y[tmpi,1]),'k',linestyle='--',\
         linewidth=3)

# No limit cycles
pars['tau']=0.02
y = integrate.odeint(pp_dyn_type2,y0,t,args=(pars,))
tmpi = np.where(t>900)
plt.plot(t[tmpi],np.transpose(y[tmpi,0]),color=[0.5,0.5,0.5],\
         linewidth=3)
plt.plot(t[tmpi],np.transpose(y[tmpi,1]),color=[0.5,0.5,0.5],\
         linewidth=3,linestyle='--')

ax.set_yscale('log')
```

SOLUTION: Model of Eco-evolutionary Dynamics

The model described in Eq. (11.6) can be written as follows:

```
def pp_dyn_evo(y,t,pars):
    N1 = y[0]
    N2 = y[1]
    P = y[2]

    N = N1+N2

    dydt = np.zeros(3)
    Q = pars['c1']*N1 + pars['c2']*N2
    dydt[0] = pars['r1']*N1*(1-N/pars['K1'])-pars['c1']*N1*P/(1+Q)
    dydt[1] = pars['r2']*N2*(1-N/pars['K2'])-pars['c2']*N2*P/(1+Q)
    dydt[2] = pars['eps']*(Q/(1+Q))*P-pars['m']*P

    return dydt
```

The following is the master script to run the full model of eco-evolutionary dynamics.

```
# Parameters
pars={}
pars['r1']=1.8 # hr^-1
pars['r2']=1.75 # hr^-1
pars['c1']=3 # ml/hr
pars['c2']=0.3 # ml/hr
pars['tau'] = 1
pars['K1']=1.8 # Cells/ml
pars['K2']=1.75 # Cells/ml
pars['eps']=0.5 # Conversion
pars['m']=0.3 # hr^-1

sfactor = 10**6

fig=plt.figure()
ax = fig.gca()
# Model and visualization
t = np.arange(20000)
y0=np.array([0.13,1.35,1.2])
y = integrate.odeint(pp_dyn_evo,y0,t,args=(pars,),rtol=10**-7,hmax=1)
N = y[:,0]+y[:,1]
plt.plot(t,N*sfactor,'k',linewidth=3,linestyle='--')
plt.plot(t,y[:,2]*sfactor,'k',linewidth=3)

plt.ylim(0,2.5*sfactor)
plt.xlim(19000, 20000)
```

```
plt.xlabel(r'Time, hrs',fontsize=18)
plt.ylabel(r'Density $ml^{-1}$',fontsize=18)

plt.legend(['total prey','Predators'],fontsize=12,loc='upper right',
           frameon=False)
ax.ticklabel_format(style='scientific',scilimits=(0,0))
plt.setp(ax.spines.values(),linewidth=2)
ax.tick_params(labelsize=14,width=2)
```

SOLUTION: Individual-Based Model Zone

We can run the model 500 times with `predator['r'] = 1` and `predator['r'] = 0.5` with the following additional parameters to get a histogram.

```
info['prey_density']=20 # Density/cm^2
prey['diffusion']=0.005 # cm^2/sec
predator['k']=0.3 # Detection
predator['f']=0.05 # Successful capture per time
predator['vel']=.1 # cm/sec
predator['handling_time']=0.0 # sec
predator['tau']=6 # Run time length, sec
```

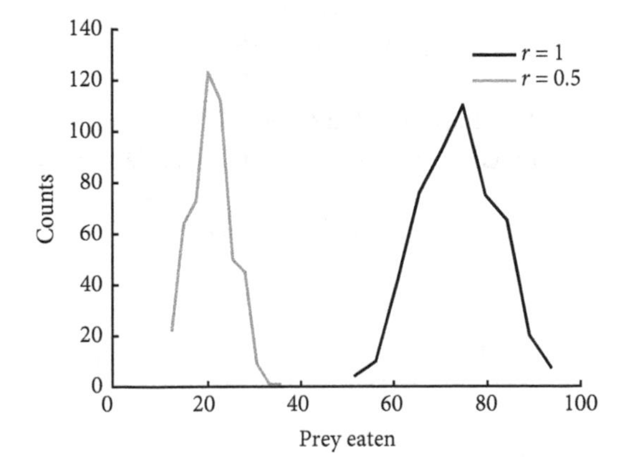

It does appear that the mean prey eaten increases by a factor of 4 when the radius increases by a factor of 2.

SOLUTION: IBM Diffusion

Vary the diffusion over 1000 repetitions with the following parameters:

```
info['prey_density']=20 # Density/cm^2
predator['r']=1.25 # Radius
predator['k']=0.3 # Detection
predator['f']=0.05 # Successful capture per time
predator['vel']=0.1 # cm/sec
predator['handling_time']=0.0 # sec
predator['tau']=6 # Run time length, sec

prey['diffusion']=np.arange(0.005,0.015,0.001)
```

This yields the following outcome:

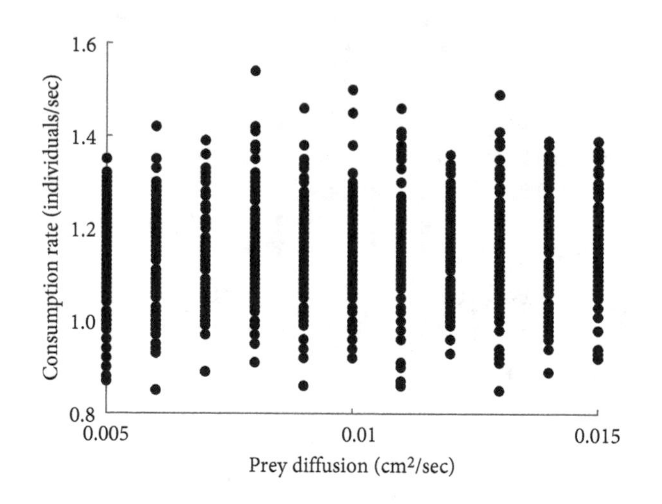

The number of prey eaten does not change as the diffusion rate varies; hence, this analysis suggests that the consumption rate is independent of the prey's diffusion rate.

SOLUTION: IBM Density

Consider the variation of the consumption rate given changes in density from 5 to 40 individuals/cm^2. We run the simulation with the following parameters:

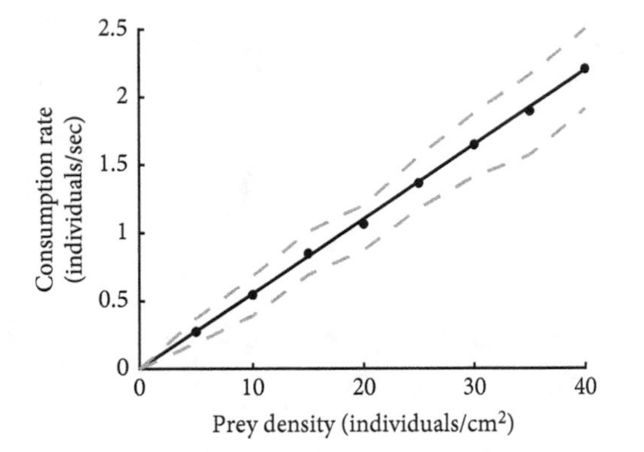


```
prey['diffusion']=0.005
predator['r']=1.25 # Radius
predator['k']=0.3 # Detection
predator['f']=0.05 # Successful capture per time
predator['vel']=0.1 # cm/sec
predator['handling_time']=0.0 # sec
predator['tau']=6 # Run time length, sec

info['prey_density']=np.arange(5,40,5) # Individuals/cm^2
```

As shown below, this is well fit by a line in log-log space with a slope of 0.99 (i.e., this is a line). The pre-factor is $0.055 \approx 0.06$, whereas the theory for a type I function response would predict a prefactor of 0.07, so there is near-quantitative agreement here.

Chapter Twelve

Outbreak Dynamics: From Prediction to Control

12.1 OUTBREAKS: FROM DETERMINISTIC MODELS TO STOCHASTIC REALIZATIONS

How do public health scientists use models of outbreaks to forecast the shape of outbreaks to come, to interpret the shape of outbreaks that have already taken place, and/or to determine how best to intervene? This laboratory provides a direct path toward building dynamic simulation approaches to support the control and prevention of infectious disease. The laboratory leverages the core ideas of the textbook to transform epidemic theory into simulation dynamics. This need remains both essential and timeless. This computational lab guide is the result of a six-year development process, from the fall of 2015 to the fall of 2021, spanning both the Ebola virus disease (EVD) outbreak in West Africa and the global COVID-19 pandemic. By the fall of 2015, it was already apparent that computational models could be used in real time to support outbreak responses. That is, relatively simple models were used to make projections, guide interventions, and estimate parameters that could connect real-world data and ongoing issues of control. The challenge of how to connect population models continues; there have been multiple EVD outbreaks since then, including one in 2018 in the Democratic Republic of Congo, with some concern that these epidemics could become endemic. Nonetheless, there is also optimism given the development of the rVSV-ZEBOV vaccine that protects against the Zaire strain (the same strain that was the causative agent of the 2014–2015 outbreak). In contrast, the role of theory and simulation in shaping the initial and sustained response to the global COVID-19 pandemic is still being written. Models have been critical in shaping responses, but the often simplified choices made have also been criticized given the need to take actions based on inference from intentionally simplified or incomplete representations of the real world.

Indeed, it is critical to keep in mind that variants of the kinds of models developed in this laboratory are used, in practice, by academic modeling groups, governmental agencies (like the WHO and CDC), and independent nongovernmental organizations to help policy makers decide how best to allocate resources, design vaccination programs, create public health campaigns, and disseminate other interventions to prevent and control disease. Indeed, one could argue that, despite its simplicity, this laboratory actually goes above and beyond some intervention-level tools. For example, the initial "Ebola Response modeling tool" developed by the CDC to provide real-time feedback and scenario response was not a program written in Python, MATLAB, or C, but instead was a Microsoft Excel spreadsheet. The tool was announced by the CDC at https://www.cdc.gov/media/releases/2014/s0923-ebola-model-Factsheet.html and is available for inspection here: https://stacks.cdc.gov/view/cdc/24900. You might wonder: why an Excel spreadsheet? Answering such a question would take time and steer us away from the pressing issues in this laboratory, but suffice it to say that the kinds of models developed here have greater flexibility but do require that one knows how to program in order to manipulate them. From a scientific perspective, there is a deeper advantage: the core equations here include feedback between different types of individuals, whether susceptible, infectious, recovered, or removed, as well as potential interventions to move individuals from one state to another (e.g., via safe burial) or to change interaction rates between individuals (e.g., via quarantine). By using a high-level programming language and not a spreadsheet, we are able to develop a suite of flexible approaches, including dynamic and stochastic models. These models are also more robust—technically. The state of modeling in response to the COVID-19 pandemic reflects a maturation of real-time response efforts: including epidemic models, model-data integration, scenario building, real-time estimators, and more. This lab can't do all of that. But it does provide the core principles to help you move beyond armchair epidemiology status (for more details, see Keeling and Rohani 2007). The methods here form the basis for state-of-the-art response models found in the rapid proliferation and interest in public health globally.

In brief, this computational laboratory will teach you how to (i) model an epidemic outbreak; (ii) explore the relationship between speed, strength, and size of an outbreak; (iii) develop a stochastic simulation of disease dynamics, flexible enough to be applied to real-world cases. This is a lot—but it builds on concepts you've already explored earlier in the book. As always, you can do it!

12.2 EPIDEMIC MODELING—FUNDAMENTALS

12.2.1 Transitions to Outbreaks

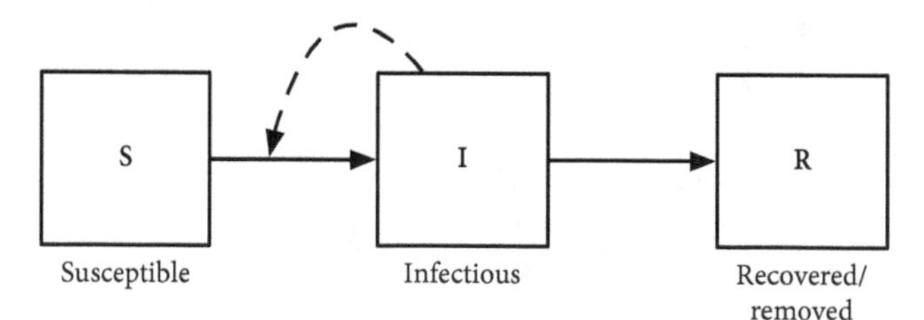

Figure 12.1: Disease dynamics represented in terms of susceptible, infectious, and recovered/removed individuals, i.e., an SIR model. Infectious individuals can come into contact with susceptible individuals, leading to new infections. Infectious individuals recover or are removed from circulation (in the case of potentially fatal diseases). These interactions form the basis for the rates and dynamics in both deterministic and stochastic versions of population-level epidemic models.

Consider an infectious disease that can spread in a population of size N. Individuals may be susceptible, infectious, or recovered/removed (Figure 12.1). We term these individuals S, I, and R, respectively, such that $S+I+R=N$. The rescaled dynamics can be written as

$$\dot{S} = -\beta SI$$

$$\dot{I} = \beta SI - \gamma I$$

$$\dot{R} = \gamma I$$

given two processes: infection at a rate βSI and recovery at a rate γI. The main text includes the full derivation of these equations and the meaning of β. Note also that variability in susceptibility can lead to fundamental changes in these equations. This model denotes the fraction of individuals of each type, i.e., $S+I+R=1$. Simulations of this model should use the following core code, beginning with the representation of the ODE:

```
def sir_model(y,t,pars):
    # function dydt = sir_model(y,t,pars)
    # SIR model

    S = y[0]
    I = y[1]

    # The model
    dSdt = -pars['beta']*S*I
    dIdt = pars['beta']*S*I-pars['gamma']*I
    dRdt = pars['gamma']*I

    dydt = [dSdt, dIdt, dRdt]
    return dydt
```

Applying this model requires a relevant choice of epidemiological parameters. Consider an infectious individual that interacts with c individuals per unit time, of which S/N are infectious, and of which a fraction p of such contacts lead to a new infectious event. In that case, we expect the rate of transmission per infectious individual to be cpS/N multiplied by the number of infectious individuals to yield $cpSI/N$ new infections per unit time. This can be written as $\beta SI/N$ given $\beta \equiv cp$.

CHALLENGE PROBLEM: Outbreak Criteria

Complete the simulation code below for the spread of an infectious disease beginning with 1 individual out of 10,000 and estimate the value of $\mathcal{R}_0$. Do you expect the disease to spread or not? Then change the value of p to 0.01. Will the disease spread? Why or why not?

```
# Main data goes here
from scipy import integrate
pars={}
pars['c'] = 20 # Contacts per unit time (days)
pars['p'] = 0.025 # Probability of infectious contact
pars['beta']=... # Transmission rate
pars['gamma']=1.0/4 # Recovery rate (1/days)
pars['basR0'] = ...
pars['N'] = 10000
```

```
pars['I0'] = 1
pars['S0'] = pars['N']-pars['I0']

# Run the model
t=np.arange(100)
y = integrate.odeint(sir_model,
        np.array([pars['S0'],pars['I0'], 0])/pars['N'],t,args=(pars,))

# Plot the results
plt.plot(t,y)
plt.xlabel('Time (days)')
plt.ylabel('Population fraction')
```

12.2.2 Speed, Strength, and Size

Once an outbreak starts, how far will it go? That is to say, how many individuals in a population will become sick? The relationship between speed, strength, and size is critical to understanding outbreak dynamics. Let's define each term. The speed can be measured in terms of the rate r of exponential growth of the outbreak, i.e., $I(t) \sim e^{rt}$. The strength can be measured in terms of the dimensionless number $\mathcal{R}_0$. This value is known as the *basic reproduction number*. It denotes the average number of new infections caused by a single infectious individual in an otherwise susceptible population. Finally, the size of the outbreak can be measured in terms of the fraction of the population infected at the culmination of the outbreak, $R(t \to \infty)$, which is equivalent to $1 - S(t \to \infty)$. The main text presents a derivation of the strength-size relationship, specifically

$$\mathcal{R}_0(S_\infty - 1) = \log(S_\infty), \tag{12.1}$$

as well as the strength-speed relationship, i.e.,

$$r = \gamma(\mathcal{R}_0 - 1). \tag{12.2}$$

Here we explore both of these systematically, gaining an intuition for these two relationships.

Strength and speed It would seem that outbreaks with higher values of $\mathcal{R}_0$ would lead to a faster increase in case counts. But that is not necessarily the case. One way to explore this is to modulate both β and γ, i.e., the transmission and recovery rates, which influence both strength and speed. To begin, let's first measure the speed in an outbreak. The code below simulates an outbreak using $\mathcal{R}_0 = 2$, $\beta = 0.5$ days^{-1}, and $\gamma = 0.25$ days^{-1}. For this combination, we expect $r = 0.25$ days^{-1}. The code below illustrates the following steps: First, simulate the model. Second, fit a line to the log-transformed infectious counts, $I(t)$. Note that you could also find the line of best fit to the cumulative case incidence, $I(t) + R(t)$. The slope of this line should be the estimated $\hat{r}$.

```
# Run the model over 10 days
t=np.arange(11)
y = integrate.odeint(sir_model,
    np.array([pars['S0'], pars['I0'], 0])/pars['N'],t,args=(pars,))

# Find the slope
p = np.polyfit(t,np.log(y[:,1]),1)

# Plot the data and overlay the best-fit exponential
fig = plt.figure()
ax = fig.gca()
plt.scatter(t,y[:,1],color='k',facecolor = 'none')
ax.set_yscale('log')
plt.plot(t,np.exp(p[0]*t+p[1]),color='r',linewidth=2)

# Use solid points for the future k
t = np.arange(30)
y = integrate.odeint(sir_model,
                     np.array([pars['S0'],pars['I0'],0])/pars['N'],
                     t,
                     args = (pars,))
tmpi = np.argwhere(t>10)
plt.scatter(t[tmpi],y[tmpi,1],color='k')
```

When implemented, this code results in the following outcome:

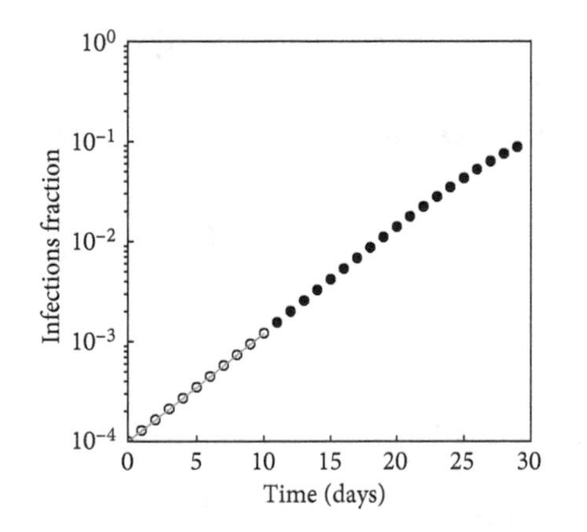

As is apparent, the fraction of infectious individuals does grow exponentially to begin, at precisely the rate expected in theory. In this case, the slope of the line is 0.2496 days^{-1} and the expected value is $r = 0.25$ days^{-1}, the difference is due to the slight depletion of susceptibles. Your next challenge is to examine the strength-speed relationship given variation in β and γ.

CHALLENGE PROBLEM: Strength-Speed Relationships

Determine the strength and speed for the following sets of disease parameters:

Transmission β	Recovery γ	Strength $\mathcal{R}_0$	Speed r
0.5	0.4	...	...
1	0.5	...	...
0.25	0.5	...	...
0.75	0.25	...	...

Strength and size An outbreak is not expected to infect all individuals. The final size of the epidemic can be measured in terms of the cumulative case count, i.e., the value of $R_\infty = 1 - S_\infty$. The next code set shows how to systematically modify the initial fraction of infected individuals. In these overlaid simulations, the initial values are marked by black circles (on the right side of each arc) and the final values by white circles (on the left of each arc). The dynamics in the phase suggest there is a critical value of $1/\mathcal{R}_0$ where an initial outbreak will not spread. This observation is the basis for findings that not all individuals in a population must be vaccinated for a vaccine to be effective at the population level! However, an ancillary finding is that the fraction of the population that should be vaccinated increases with the strength of the disease, i.e., measured in terms of $\mathcal{R}_0$.

```
# Modify the initial values
pars['N'] = 10000
pars['S0_range'] = np.array([0.6, 0.7, 0.8, 0.9, 0.999])
pars['I0_range'] = 1/pars['N']*np.ones(5)
pars['R0_range'] = 1 - pars['S0_range'] - pars['I0_range']

# Run the model
fig = plt.figure()
ax = fig.gca()
plt.xlim([0,1])
plt.ylim([0,0.25])
for i in range(len(pars['S0_range'])):
    t = np.arange(200)
    y = integrate.odeint(sir_model,
        [pars['S0_range'][i], pars['I0_range'][i],\
        pars['R0_range'][i]], t,args=(pars,))
    plt.plot(y[:,0],y[:,1],color='k',linewidth=3)
    plt.scatter(y[-1,0],y[-1,1],color='r',s=100)
    plt.scatter(y[0,0],y[0,1],color='k',s=100)
```

```
# Show the excluded regime
from matplotlib.patches import Polygon
verts = [(1,0), (1,0.25), (0.75,0.25)]
poly = Polygon(verts, facecolor=[0.8,0.8,0.8])
ax.add_patch(poly)
```

Running this code yields the following trajectories, where the shaded region denotes inaccessible parts of phase space given $S + I \leq 1$.

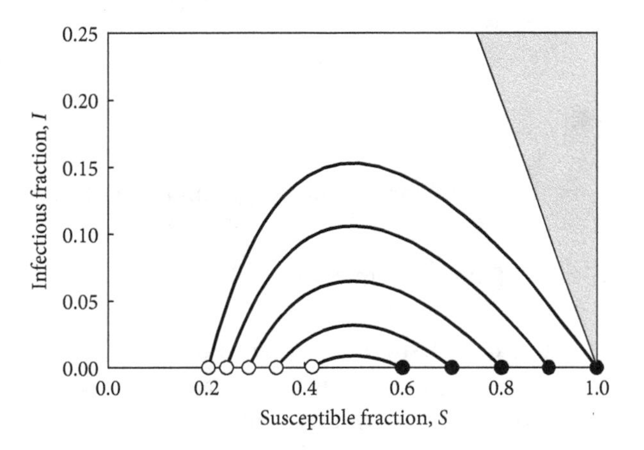

This example of visualizing outbreak dynamics in the phase plane also provides the basis for the next challenge problem.

CHALLENGE PROBLEM: Strength-Size Relationships in Phase Space

First, fix the value of the recovery rate to $\gamma = 0.25$, then modulate the transmission rate from $\beta = 0.25$ to $\beta = 2.5$. Assume that, initially, 80% of the population is susceptible and 20% is recovered. In this case, does the disease always spread? Why or why not? In addition, find the outbreak size and show how it relates to $\mathcal{R}_{eff}$—the effective reproduction number given the initial susceptible population.

12.2.3 From Outbreaks to endemics

The SIR model, with rapid loss of immunity, can be written in terms of an SI model:

$$\dot{S} = -\beta SI + \gamma I$$
$$\dot{I} = \beta SI - \gamma I$$

This model assumes that there is no immunity. Instead, recovered individuals are immediately susceptible. As a result, the dynamics include a new long-term possibility: an endemic disease state where $S^* = \frac{1}{\mathcal{R}_0}$ and $I^* = \frac{\mathcal{R}_0 - 1}{\mathcal{R}_0}$. This SI model can be coded as follows:

```
def si_model(y,t,pars):
    # function dydt = si_model(y,t,pars)
    # SI Model

    S = y[0]
    I = y[1]

    # The model
    dSdt = -pars['beta']*S*I + pars['gamma']*I
    dIdt = pars['beta']*S*I - pars['gamma']*I

    dydt = [dSdt, dIdt]
    return dydt
```

As should be apparent, there is only an endemic state with $I^* > 0$ when $\mathcal{R}_0 > 1$.

CHALLENGE PROBLEM: From Epidemic Outbreaks to Endemics

Using $\beta = 0.3$ and $\gamma = 0.25$ days^{-1}, show the convergence of the dynamics to an endemic state, i.e., from an outbreak to persistent infectious cases. Then compare and contrast this with dynamics in which immunity is permanent. For these parameters, simulate the dynamics over a 1 year = 365-day period.

12.3 STOCHASTIC EPIDEMICS

12.3.1 Definition of the model

Stochastic realizations of the SIR model are simulated using the Gillespie framework, given the "reaction" events in the following table:

Process	Reaction	Rate/probability
Infection	$S + I \rightarrow 2I$	$r_1 = \beta_I S \frac{I}{N}$
End of infectiousness (recovery)	$I \rightarrow R$	$r_2 = \gamma I$

These processes denote transitions among individuals who are susceptible (S), infectious (I), and recovered/removed (R). The total population is fixed at $N = S + I + R$. Epidemics are initiated with one infectious individual in an otherwise susceptible population. Mathematically, the initial state is $\mathbf{y} = (N_0 - 1, 1, 0)$ at $t = 0$, where the ordering of terms is susceptible, infectious, and recovered. The total rate of outbreak-associated events is $r_{tot} = \sum_{i=1}^{2} r_i$. The time until the next event is determined randomly such that $\delta t \sim \frac{-\log \chi}{r_{tot}}$ where χ is a uniformly distributed number between 0 and 1. In this way, the time between events follows an exponential distribution with rate r_{tot}. Then the probability of each event is r_i / r_{tot}. The process stops when $I(t) = 0$, as in that case $r_{tot} = 0$. After selecting an event and updating the discrete number of individuals, the reaction rates are recalculated and the process continues. The same framework can be extended to include other classes (e.g., exposed individuals) as well as other kinds of processes (e.g., postdeath transmission in the case of EVD). Trajectories are complete when the epidemic dies out because there are no more infectious individuals. In the present context, we are interested in those trajectories that do not die out before the end of the simulation time.

12.3.2 Simulating a stochastic outbreak

In Chapter 3, we developed Gillespie models of gene expression. The same principles can be applied here. Here your objective is twofold: first to modify the code below to simulate a stochastic epidemic, and then to simulate and compare results from stochastic simulations to your deterministic trajectories.

CHALLENGE PROBLEM: Stochastic SIR Model

Modify the following code to implement a stochastic SIR model, including both transmission and recovery events, in a population of size *N*.

```
def stochsim_SIR(y0,trange,pars):
    # function [t,y] = stochsim_SIR(y0,trange,pars)

    # Simulates a SIR model via the Gillespie algorithm
    # from t0 to tf in trange given initial
    # conditions in y0 = [S0 I0 R0] and parameters
    # in pars. Returns time and values

    # Conditions
    t0 = trange[0]
    tf = trange[1]
    t = [t0]
    # We start y as a list of [S,I,R], where each triplet represents
    # a time point
    y = [y0]
    tcur = t0
    ycur = list(y0)
    ind = 0
    # Model
    while tcur<tf:
        # Check to see if there is an infection
        if ycur[1]==0:
            ind = ind+1
            t.append(tf)
            y.append(ycur)
            break
        # Rates
        infrate = ...
        recrate = ...
```

```
        totrate = infrate + recrate
        dt = -1/totrate * np.log(np.random.uniform())
        tcur = tcur+dt
        # Event type
        if np.random.uniform()<(infrate/totrate): # Infection
            ...
            ...
        else: # Recovery
            ...
            ...
        ind = ind+1
        t.append(tcur)
        y.append(list(ycur))
    # Now recast y as a numpy array to make it easier to work with
    y = np.array(y)
    return [t,y]
```

Once you have a working code, simulate it using the following parameters, including an initial seed of 10 infectious individuals. This is usually enough to ensure the disease spreads. The following parameters should provide enough context and information to intialize the simulation.

```
pars['c']=20
pars['p']=0.025
pars['beta']=pars['c']*pars['p']
pars['gamma'] = 0.25
pars['tf']=60
pars['basR0']=pars['beta']/pars['gamma']
pars['N']=1000
pars['I0']=10
pars['S0']=pars['N']-pars['I0']
```

Comparing directly to the SIR model should yield the following result. Here both the stochastic (dashed line) and deterministic (solid line) trajectories are plotted against each other.

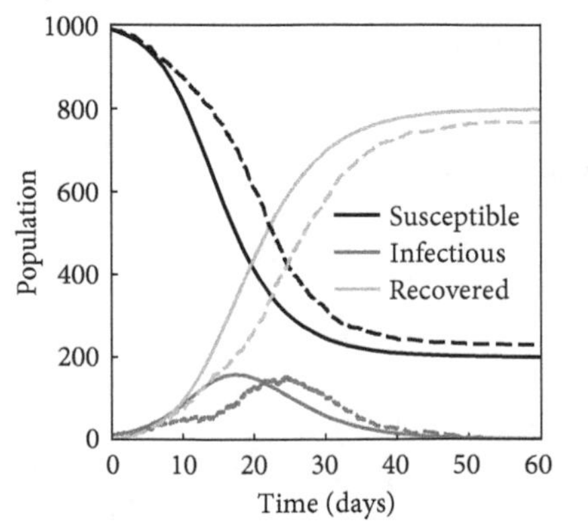

As is apparent, there can be many differences between a deterministic model and a stochastic trajectory. Understanding that difference is at the very heart of inferring and predicting disease trajectories in real systems. If you have time, you could even try to estimate the speed of the outbreak for a set of trajectories. You will find that there is variability even without considering observational noise or process noise, and that such variation can be substantial at the outset of an outbreak when estimates are most needed!

Finally, it may seem difficult to compare the models, but it is easier than it seems. The key is to run both models but then recall that the ODE model dynamics are in terms of fractions, which must be rescaled. Here is how to do it:

```
# Run the ODE model
t=np.linspace(0,pars['tf'])
y = integrate.odeint(sir_model,
            np.array([pars['S0'],pars['I0'],0])/pars['N'],
                   t,
                   args=(pars,))
plt.plot(t,pars['N']*y[:,0],linewidth=3,color='b')
plt.plot(t,pars['N']*y[:,1],linewidth=3,color='r')
plt.plot(t,pars['N']*y[:,2],linewidth=3,color='g')

# Run the stochastic model
[tsim,ysim] = stochsim_SIR([pars['S0'],pars['I0'],0],[0,pars['tf']],pars)
plt.plot(tsim,ysim[:,0],linewidth=3,color='b',linestyle = '--')
plt.plot(tsim,ysim[:,1],linewidth=3,color='r',linestyle = '--')
plt.plot(tsim,ysim[:,2],linewidth=3,color='g',linestyle = '--')
```

The link between stochastic and deterministic models is of particular interest at the outset of an epidemic. In these circumstances, diseases may spread in ways that do not resemble typical exponential dynamics until they reach "liftoff." The consequences are that the disease may already be far more pervasive than expected by the time reliable estimates of speed and strength are made. Moreover, even estimates of speed and strength will be less certain, precisely because of the impact of stochasticity.

SOLUTIONS TO CHALLENGE PROBLEMS

SOLUTION: Outbreak Criteria

The missing components are shown below.

```
pars['beta']=pars['c']*pars['p'] # Transmission rate
pars['basR0'] = pars['beta']/pars['gamma']
```

These parameters describe an outbreak for the SIR model given $\mathcal{R}_0 = 2$, $\beta = 0.5$ days^{-1}, $\gamma = 0.25$ days^{-1}. As a result, the simulation of the outbreak looks as follows:

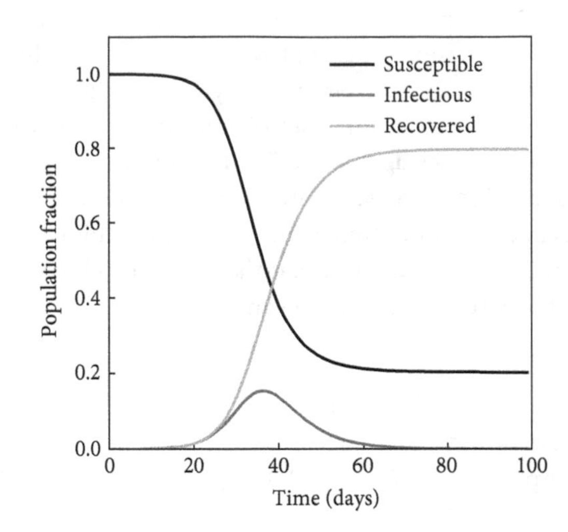

In this case, the outbreak occurs because $\mathcal{R}_0 > 1$. However, when $p = 0.01$, then $\beta = 0.2$ and $\mathcal{R}_0 = 0.8$, which is less than 1. A simulation (results not shown) leads to the recovery of infectious individuals without a net increase in the number of cases.

SOLUTION: Strength-Speed Relationships

The solution proceeds as before, by simulating the dynamics, fitting an exponential, and then comparing to the theoretically expected value. Notably, the third case has $\mathcal{R}_0 < 1$, which means the speed is negative, i.e., the number of cases declines over time. Altogether, the images look as follows:

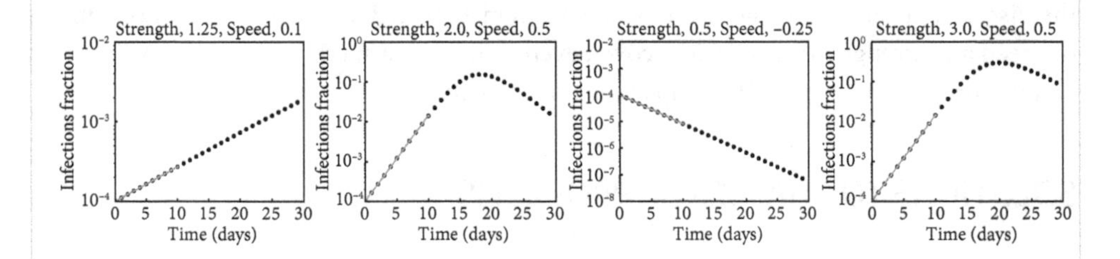

The estimated values of strength and speed are given below.

Transmission β	Recovery γ	Strength $\mathcal{R}_0$	Speed r
0.5	0.4	1.25	0.1
1	0.5	2	0.5
0.25	0.5	0.5	−0.25
0.75	0.25	3	0.5

SOLUTION: Strength-Size Relationships in Phase Space

The effective reproduction number $\mathcal{R}_{eff} = \frac{\beta S_0}{\gamma}$, which ranges from 0.8 to 8 in this case. Hence, even though $\beta > \gamma$, there will not be an outbreak until $\beta > \gamma/S_0$, or $\beta_c = 5/16 = 0.3125$. This is apparent in the plots below (left, phase space; right, relationship between final size and β). As noted in the right plot, the ratio of β/γ must exceed 1/0.8, or 1.25, for the outbreak to initiate. Analytical derivations of the strength-size relationship given a partially susceptible population are presented in the main text.

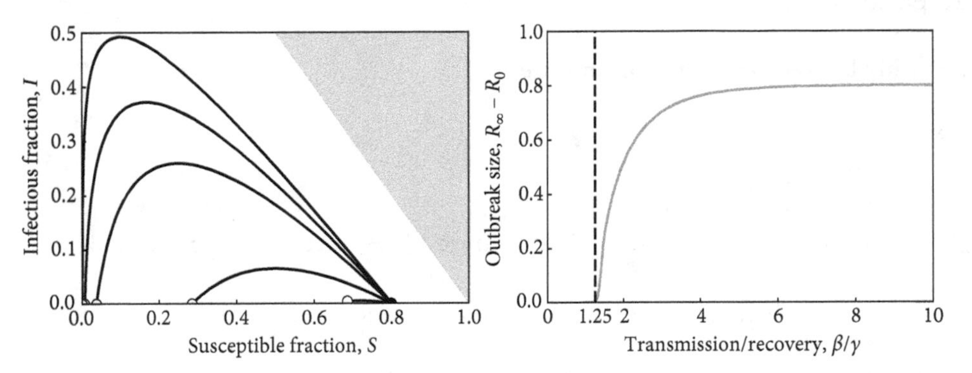

The code to generate the phase plots is given below. Note that the labels are not included and should be added to enhance interpretation and readability.

SI phase dynamics

```
pars['N']  = 10000
pars['S0']  = 0.8
pars['I0']  = 1/pars['N']
pars['R0']  = 1 - pars['S0'] - pars['I0']
pars['beta_range'] = [0.25 0.3 0.35 0.5 1 1.5 2.5]

# Run the model
fig = plt.figure(figsize=(7,5))
ax = fig.gca()
plt.xlim([0,1])
plt.ylim([0,0.5])
for i in range(len(pars['beta_range'])):
    pars['beta'] = pars['beta_range'][i]
    pars['basR0'] = pars['beta']/pars['gamma']*pars['S0']
    t = np.arange(0,200,0.1)
    y = integrate.odeint(sir_model,
                         np.array([pars['S0'], pars['I0'],\
                         pars['R0']]), t, pars,))
    plt.plot(y[:,0],y[:,1],color='k',linewidth=3)
    plt.plot(y[-1,0],y[-1,1],marker='o',markerfacecolor='w',
            markeredgecolor='k',markersize=10)
```

```
    plt.plot(y[0,0],y[0,1],marker='o',markerfacecolor='k',
             markeredgecolor='k',markersize=10)

# Show the excluded regime
from matplotlib.patches import Polygon
verts = [(1,0), (0.5,0.5), (1,0.5)]
poly = Polygon(verts, facecolor=[0.8,0.8,0.8])
ax.add_patch(poly)
```

The relationship between transmission and size

```
pars['N'] = 10000
pars['S0_range'] = 0.8
pars['I0_range'] = 1/pars['N']
pars['R0_range'] = 1 - pars['S0_range'] - pars['I0_range']

R_end = np.zeros(np.shape(pars['beta_range']))
# Run the model
fig = plt.figure(figsize=(7,5))
ax = fig.gca()

for i in range(len(pars['beta_range'])):
    pars['beta'] = pars['beta_range'][i]
    pars['basR0'] = pars['beta']/pars['gamma']*pars['S0_range']
    t = np.arange(0,200,0.1)
    y = integrate.odeint(sir_model, np.array([pars['S0_range'],
        pars['I0_range'], pars['R0_range']]), t, args=(pars,))
    R_end[i]=y[-1,2]

plt.plot(pars['beta_range']/pars['gamma'],R_end-pars['R0_range'],\
         linewidth=3, color=[0.5,0.5,0.5])
plt.plot([1.25,1.25],[0,1],linestyle='--',color='k',linewidth=3)
plt.xlabel(r'Transmission/recovery, $\beta/\gamma$',fontsize=20)
plt.ylabel(r'Outbreak size, $R_\infty-R_0$',fontsize=20)
plt.xticks([0,1.25,2,4,6,8,10])
ax.set_xticklabels(['0','1.25','2','4','6','8','10'])
```

SOLUTION: From Epidemic Outbreaks to Endemics

The contrast between the SIR model and the SI model can be seen in the plots of the contrasting outcomes shown below. In both cases, there is an outbreak. However, with $\mathcal{R}_0 = 1.2$, we expect 83.3% of the population to be susceptible at equilibrium and therefore 16.7% to be infected—precisely as observed.

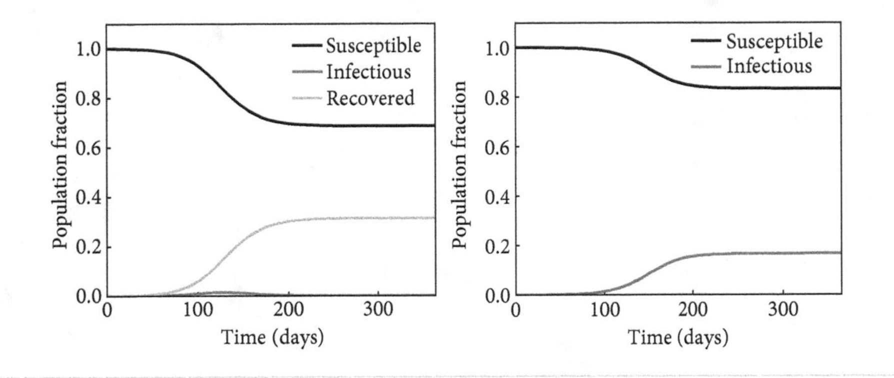

SOLUTION: Stochastic SIR Model

There are two key components to modify. First, the underlying transition rates must be modified. Those include the infection and recovery rates. For a generalized model, there may be other rates as well. Here is the correct code:

```
# Rates
infrate = pars['beta']*ycur[0]*ycur[1]/pars['N']
recrate = pars['gamma']*ycur[1]
# Event type
if np.random.uniform()<(infrate/totrate): # Infection
    ycur[0] = ycur[0]-1
    ycur[1] = ycur[1]+1
else: # Recovery
    ycur[1] = ycur[1]-1
    ycur[2] = ycur[2]+1
```

The key here is that there are multiple state variables to change, and they go up/down by 1 depending on whether there is a new infection or a recovery event.

Part IV

The Future of Ecosystems

Chapter Thirteen

Ecosystems: Chaos, Tipping Points, and Catastrophes

13.1 MODELING COMPLEXITY: AN ENABLING VIEW

Ecosystem dynamics arise from the interactions of heterogeneous agents, a combination of biotic and abiotic factors, and often strong influences of stochasticity. Such a combination of factors would tend to lead the quantitative study of particular ecosystems toward larger and larger models, with ever more interactions, feedback, parameters, and so on. Yet, as described in the textbook chapter, features of ecosystem dynamics can often transcend the particular details. What then is the right balance to strike? If made too simplistic, models become no more than metaphors or, worse, objects meant to be admired, not falsified. If too complex, then it becomes difficult, if not impossible, to identify the salient factors that govern key phenomena other than in a statistical insight. This lab guide attempts to strike a balance by contextualizing and motivating the biology (and physics, when necessary) of model choices so as to connect dynamical models with mechanisms. Nonetheless, this kind of work, when adopted at ecosystem scales, poses certain challenges.

Precisely so, this final computational laboratory departs from the rule in the spirit of the adage "the exception that proves the rule." The exception is precisely because models of ecosystems do, in fact, tend to have many components, including population abundances, intraspecies variants, nutrient inputs, patches/habitats for colonization, and other physical drivers that couple to ecology. They are the closest manifestation of the warning by Rosenblueth and Wiener (1945) on model building: "The best material model of a cat is another, or preferably the same, cat." Yet what has given life to these models are a few simple ideas; for example: small changes in conditions can lead to large changes in outcomes; states—when entered—can persist for far longer and under more extreme conditions than it took to enter them; certain types of feedback can lead to runaway, positive feedback events that can ultimately lead to catastrophe.

In essence, the point of the models in this laboratory is to identify the dynamical ingredients that lead to the emergence of chaos, catastrophes, and alternative stable states in communities and ecosystems. Because we are focusing on ingredients, each of the model sections is relatively small and builds in complexity, motivated by recent models connected to observational datasets that are only slightly more involved.

First, a simplified model of logistic growth is presented using discrete time steps—representative of seasons or growing periods. This model is related to some of the earliest models in the book, with a major exception: even this simple model can exhibit chaos (May 1979). Second, the laboratory switches back to a continuous frame, again turning to the logistic model, with an exception: namely, that the model assumes that the population reshapes its environment. By reshaping the environment, it is possible that a population can increase its growth rate, but in certain cases this feedback ends not in a steady state but in a catastrophe (Cohen 1995). Finally, we look at a model of a looming catastrophe: the Earth system. In doing so, we focus on a particular low-dimensional model of climate, shaped by phytoplankton where, in some sense, their actions are implicit. The model reveals, as you will see, that if we think about the Earth as a dynamical system, then like neurons and cardiac cells, it is possible that a rapid influx of organic carbon can lead to excitations and even long trajectories far from the equilibrium state (Rothman 2017, 2019).

13.2 SMALL DIFFERENCES, BIG EFFECTS

How do small variations in initial conditions or parameters affect the long-term dynamics of populations, communities, and ecosystems? For linear systems, the answer is trivial: small variations have small effects. However, for nonlinear systems, the answer depends on the feedback of the system itself. One might expect that the variation could range from small to large, depending on the system. What is more interesting is that, for nonlinear dynamical systems, the answer to this question can be extreme. In some cases, any variation, however small, will eventually lead to a complete loss of predictability/correspondence between two otherwise identical systems. This is the butterfly effect in a nutshell: that a seemingly imperceptible change in the system conditions can lead to a completely different outcome. Let's see how this works in perhaps the simplest model that exhibits this phenomenon—the discrete logistic model.

13.2.1 Logistic growth versus logistic maps

The very earliest models from Chapters 1 and 2 of the textbook and this laboratory guide focused on problems related to bacterial reproduction. Successive divisions could be described as a continuous process, as

$$\frac{dN}{dt} = rN \tag{13.1}$$

where r is the growth rate of the population, such that $N(t) = N_0 e^{rt}$ given an initial abundance of N_0 at time $t = 0$. Notably, given that $N(t) = 2N_0$ when $e^{rt} = 2$, then the "doubling" time can be identified as

$$\tau = \frac{\log 2}{r}. \tag{13.2}$$

Yet bacteria (or any other species for that matter) cannot grow indefinitely. Instead, there will be limits. One way to impose limits is to denote the growth of the population via a logistic model such that

$$\frac{dN}{dt} = rN\left(1 - \frac{N}{K}\right). \tag{13.3}$$

Here, as long as $r > 0$ and $N_0 > 0$, there can only be one possible outcome: the system will converge rapidly to $N = K$, on the time scale of order $1/r$ (i.e., similar to the doubling time).

Yet not all processes are best described by continuous dynamics, particularly those with growing seasons or other staged/structured periods of growth and decay. Given the nature of doubling, the following discrete growth model would seem equivalent to the continuous model:

$$N_{t+1} = RN_t \tag{13.4}$$

where R denotes a discrete growth factor and N_t denotes the population at time t. The solution is $N_t = N_0R^t$, where $\log R$ is the discrete exponential growth factor. As such, as long as $R > 1$, the system will grow in population. But nothing grows forever. Hence, first extend this model to include a logistic feedback term, i.e.,

$$N_{t+1} = RN_t\left(1 - \frac{N_t}{K}\right). \tag{13.5}$$

To do so, enter in the following model code given $R = 1.5$ (Figure 13.1) or develop one on your own. Either way, overlay the results of two sets of initial conditions—one relatively large and the other relatively small—to visualize the dynamics as follows. Once you have a working code, try out the next challenge problem.

```
import numpy as np
import matplotlib.pyplot as plt

# Parameters
pars={}
pars['R']=1.5
pars['K']=100
pars['tmax']=20
pars['dt']=1

# Simulation
t = np.arange(0,pars['tmax']+pars['dt'],pars['dt'])
N = np.zeros(np.shape(t))
N[0]=1
for i in range(len(t)-1):
    N[i+1]=pars['R']*N[i]*(1-N[i]/pars['K'])

# Visualization
plt.plot(t,N,c='k',marker='o',markerfacecolor='w')
plt.xlabel('Time, t',fontsize=20)
plt.ylabel('Population, N',fontsize=20)
```

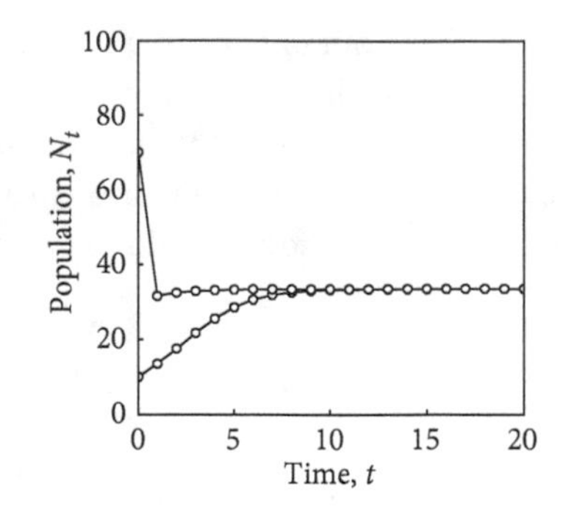

Figure 13.1: Logistic model with $R = 1.5$ and $K = 100$.

CHALLENGE PROBLEM: Finding the Equilibria

Utilize the discrete logistic model and find the equilibria for the cases $R = 1.2$, 1.5, and 1.8. Visualize the dynamics to show that simulations agree with the analytical predictions.

This logistic model can be normalized by rewriting the model in terms of the fractional abundance $x_t = N_t/K$ such that

$$x_{t+1} = Rx_t\,(1 - x_t). \tag{13.6}$$

In this model, the fractional abundance is constrained such that $0 \leq x_t \leq 1$. This model is the basis for the next section of analysis.

13.2.2 Bifurcations and oscillations in logistic maps

The discrete logistic model denotes a *map* between the current state of the system x_t and the next state of the system x_{t+1}. Formally, this is $f: X_t \rightarrow X_{t+1}$. To get started, use the following code to visualize the map. As you will see, the system has 0 population when initialized at either $x = 0$ or $x = 1$, and has a maximal population level at the next time period when initialized at $x = 0.5$, illustrated in the code snippet below.

```
# Parameters
pars={}
pars['R']=3.2
x0 = np.arange(0,1.001,0.001)

# The model
disc_logistic = lambda x: pars['R']*x*(1-x)

#Plot the map
fig = plt.figure(figsize=(5,5))
ax = fig.gca()
plt.plot(x0,disc_logistic(x0),c='k',linewidth=3)
```

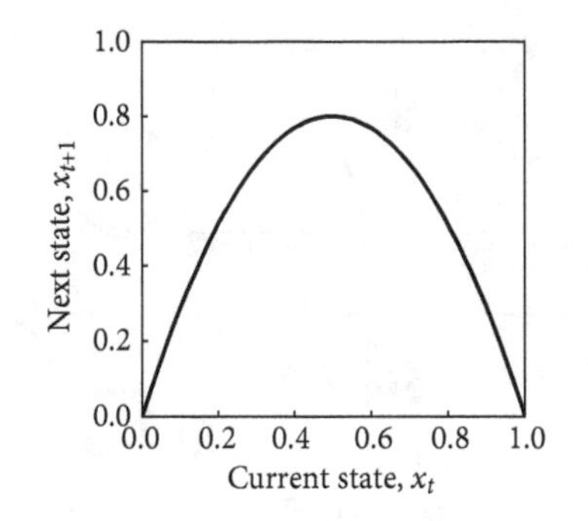

Figure 13.2: Discrete logistic model map with $R = 3.2$.

The logistic map (shown in Figure 13.2) can also be used as the basis for understanding long-term dynamics. Consider the following procedure. First, start with the current state, x_t. Then use the map to identify the next state, x_{t+1}. If you use the map a second time, it is possible to identify the next-next state, i.e., x_{t+2}. Visually, this amounts to "cob-webbing"—remapping the y value back to the x value and then mapping again and again. Depending on the value of R, the system may converge to a single equilibrium state x^*, to multiple values (denoting oscillations in the dynamics), or to a series that is completely unpredictable over the long term (chaos!!!). The next challenge problem is your chance to write a cobwebbing code and then explore the dynamics in the long term. Are long-term dynamics predictable (or not) despite the fact that the near-term dynamics are completely predictable?

CHALLENGE PROBLEM: Cobwebbing Dynamics

Write code that simulates the discrete logistic model and overlays the results directly on the map. Use a duration of 20 time points (think of them as growing seasons). Do this for three values of R, i.e., $R = 2.8$, $R = 3.2$, and $R = 3.9$. If the code is working, it should look like this:

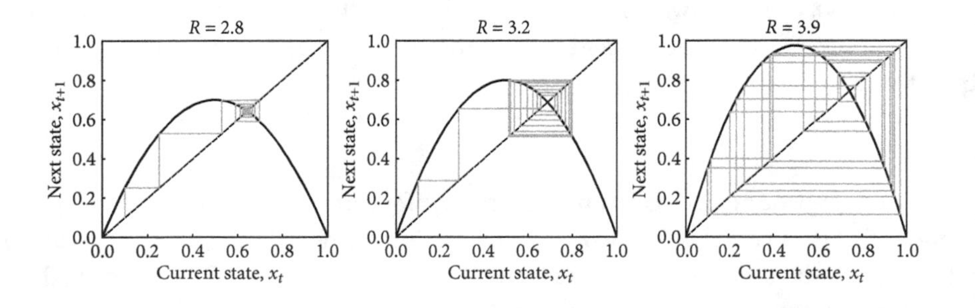

Hint: The key insight here is to think of cobwebs as combinations of two-line segments. These two-line segments form brackets connecting three sets of points: (i) (x_t, x_t), (ii) (x_t, x_{t+1}), and (iii) (x_{t+1}, x_{t+1}). Repeating again and again should yield a visualization of the dynamics that can be overlaid with the map (as shown above). This can be seen in action in the sequence of snapshots shown below for the case of $R = 3.9$—a form of cobweb animation beyond the scope of this laboratory.

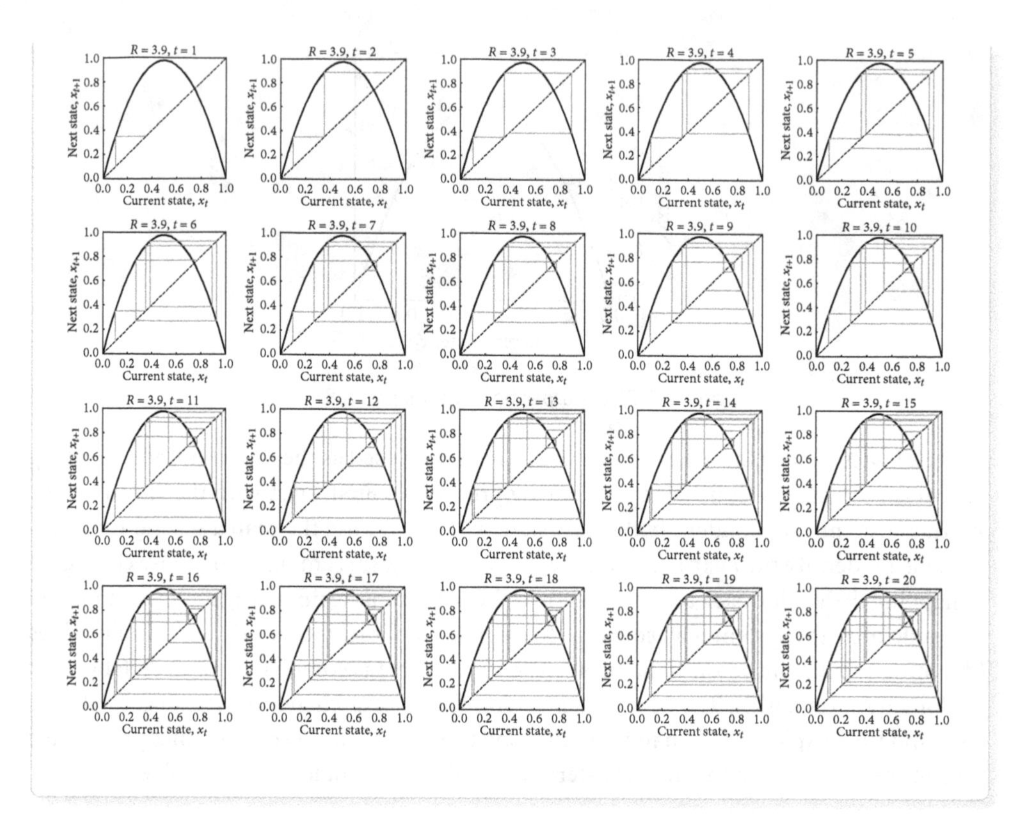

13.2.3 Evidence of long-term divergence of trajectories despite short-term predictability

How much will two systems differ in the long term even when they start off at nearly the same initial condition? The answer to this question requires significant elaboration. But, in practice, it is possible to measure divergence for the discrete logistic model (and indeed for other systems). Consider two trajectories, x_n and x'_n, which are initialized at the initial conditions, x_0 and $x_0 \pm \delta_0$. The divergence can be measured as the absolute difference between trajectories:

$$\delta x(n) \equiv |x_n - x'_n|. \tag{13.7}$$

The discrete logistic model has an unusual property for certain values of R, namely, that trajectories that start very close to each other become increasingly divergent at an exponential rate! This can be seen visually by contrasting the dynamics when $R = 2.8$ versus that of $R = 3.9$. Using the code below, generate a pair of time series separated by a small deviation, e.g., 0.0001. Then calculate the divergence and note whether it increases or decreases. This code is the basis for further exploration of the fact that the divergence doesn't just increase or decrease, but does so exponentially (as is evident by the nearly linear nature of the divergence with time when plotted on `semilogy` axes). The code below provides a solid basis for the exercise:

```
# Initialize the parameters
pars={}
pars['R']=2.8
pars['tmax']=25
pars['x0']= [0.1, 0.10001]

# Simulations
x=np.zeros((pars['tmax']+1,len(pars['x0'])))
for i in range(len(pars['x0'])):
    x[0,i]=pars['x0'][i]
    for j in range(pars['tmax']):
        x[j+1,i]=pars['R']*x[j,i]*(1-x[j,i])
t=np.arange(pars['tmax']+1)

# Divergence
deltax = np.abs(x[:,0]-x[:,1])

# Plot divergence
fig, axes = plt.subplots(1,2,figsize=(16,5))
ax = axes[1]
ax.plot(t,deltax,'k',linewidth=3)
ax.set_yscale('log')
ax.set_xlabel('Time, $t$',fontsize=20)
ax.set_ylabel(r'Divergence, $\delta x(n)$',fontsize=20)
# Plot the states
ax = axes[0]
ax.plot(t,x[:,0],c='k',linewidth=2,
        marker='o',
        markerfacecolor='w',
        markersize=12)
ax.plot(t,x[:,1],linestyle='',
      marker='d',markerfacecolor=[0.5,0.5,0.5],markeredgecolor='k',
        markersize=10)
ax.legend([r'$x_n$',r"$x_n'$"],fontsize=14,loc='lower right')
ax.set_title(r'$R={R}$'.format(R=pars['R']),fontsize=20)
ax.set_xlabel(r'Time, $t$',fontsize=20)
ax.set_ylabel(r'State, $x_n$',fontsize=20)
```

If you decide to also plot the trajectories along with the divergence, you will find the following outcomes:

As is evident, trajectories for the case of $R = 2.8$ converge rapidly; indeed, it is not even possible to see the difference in the trajectory plot. In contrast, the trajectory divergence reaches nearly the limit of the system range after 20 iterations in the case of $R = 3.9$. This exponential increase in divergence is a hallmark of chaos—a topic for postlaboratory work.

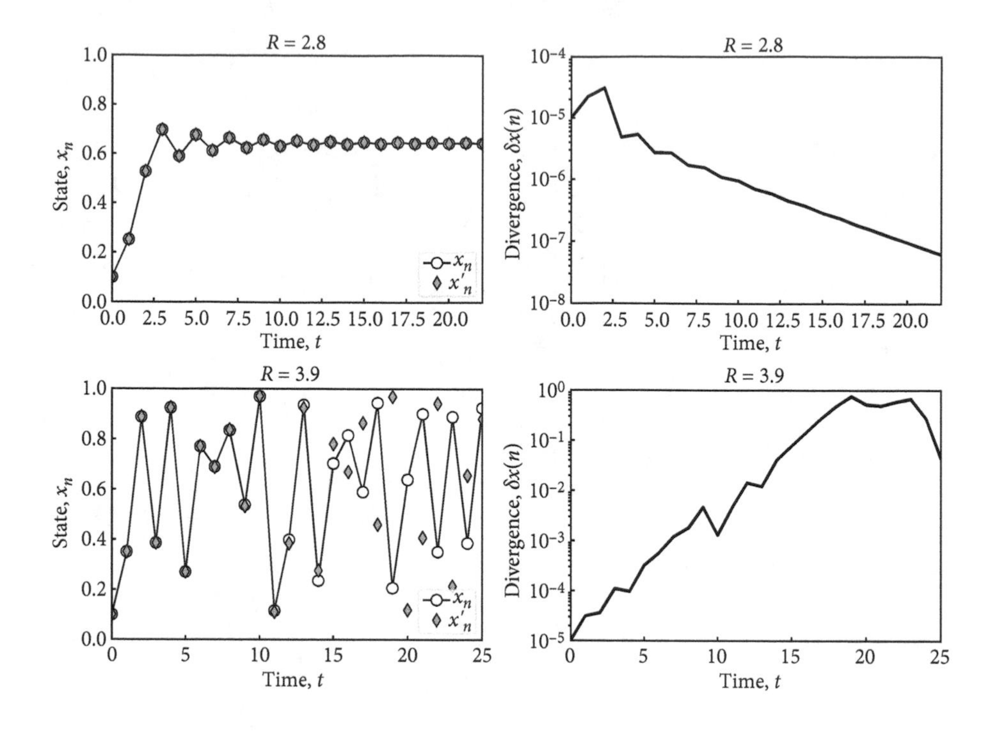

13.3 EXPLOSIVE GROWTH AND POPULATION CATASTROPHES

13.3.1 Preamble—logistic growth model

The logistic model is one of the simplest in the study of living systems. It describes the initial exponential growth of a population at a rate r that eventually reaches a saturating carrying capacity K, i.e.,

$$\frac{\mathrm{d}N}{\mathrm{d}t} = rN\left(1 - \frac{N}{K}\right). \tag{13.8}$$

The solution to this model depends on r, K, and N_0, the population abundance at $t = 0$. Given that this is the last laboratory in this guide, it is worth asking yourself: can you start with an entirely blank file, simulate the logistic model, and confirm that your answers are indeed correct—without checking with a friend or Google? For an experienced coder, this is a straightforward task. But these labs are intended to be used by scientists and students, including those with limited coding experience. Hence, to the extent possible, try to do this next challenge problem without notes and see how far you can go.

CHALLENGE PROBLEM: Continuous Logistic Model from Scratch

Write code that simulates the logistic model, over a period of 0 to 100 days, assuming a population that begins at a level $N_0 = 50$, given a maximum growth rate $r = 0.1$ and carrying capacity $K = 10^5$.

Important note: The solution to this problem has already been documented at the very outset of this lab guide. But that was a long time ago. In many courses, the same question is asked as a pre- and postexam to evaluate learning (and retention). Consider this your own benchmark.

13.3.2 Malthus-Condorcet model

The logistic model represents the impact of *negative* density dependence on the growth of a particular population. But what if the density were to have a stimulatory effect, rather than an inhibitory effect, on growth? Such a model can be considered by generalizing the logistic model as follows:

$$\frac{dN(t)}{dt} = \rho N(t)\,(K(t) - N(t)) \tag{13.9}$$

$$\frac{dK(t)}{dt} = c\frac{dN(t)}{dt} \tag{13.10}$$

where $N(t)$ is the population, $K(t)$ is the carrying capacity, ρ is a scaled growth rate, and c is the Condorcet parameter. The Condorcet parameter controls the extent to which the carrying capacity increases along with the population. The simulation code for this model is easily adapted from the logistic growth model, i.e.,

```
def condorcet_model(y,t,pars):
    # Malthus-Condorcet model dynamics
    # dNdt = rho*N*(K-N)
    # dKdt = c*dNdt
    # Initialize
    N=y[0]
    K=y[1]

    # Find derivatives
    dNdt = pars['rho']*N*(K-N)
    dKdt = pars['c']*dNdt
    return np.array([dNdt, dKdt])
```

We refer to this as the Malthus-Condorcet model; see main text for more details.

As explained in the textbook, the carrying capacity increases at a rate equal to that of the rate of change of the population modified by the Condorcet parameter, c. In addition, when $N \ll K$, then the growth rate is exponential with a rate of ρK. This model can be mapped to the logistic model when $c = 0$. However, three regimes are relevant for some form of positive feedback, $c > 0$: (i) $c < 1$; (ii) $c = 1$; (iii) $c > 1$. The value of c controls the extent to which the capacity of the system to grow increases with the population. It is worthwhile to explore the dynamics using the following parameters and initial conditions: $\rho = 10^{-4}$, $K_0 = 500$, $N_0 = 50$, albeit given $c = 0.5$, $c = 1$, and $c = 2$.

These three models all start with the same initial conditions. If K had been constant, then the maximum growth rate would be $\rho K_0 = 0.05$, setting a time scale of approximately 20. But

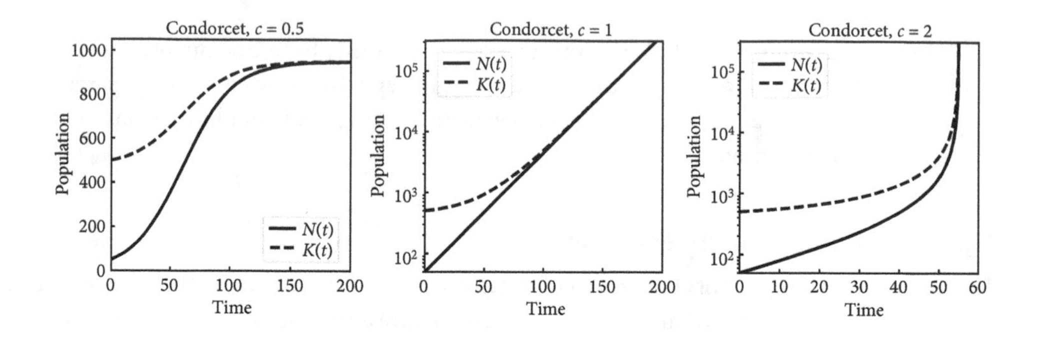

the carrying capacity is not constant. Instead, in each case the carrying capacity increases, leading to a qualitative change in dynamics. You should recapitulate these findings while observing a few salient results. First, when $c = 0.5$, the population dynamics appear to be logistic, but in fact the population saturates at a final carrying capacity nearly double that initially. Next, when $c = 1$, the system grows exponentially after an initial transient period, because the population continues to increase its capacity for growth—as it grows. At long times, the dynamics converge to $N(t) \sim Ae^{\tilde{r}t}$, where A is a constant. It is worth finding the effective exponential growth rate and comparing it to the theoretical expectation: $\rho(K(0) - N(0))$.

When $c = 2$, the system grows faster than exponentially—note the positive curvature on the log-scaled axis. Yet, in running the code, there is another kind of problem: any standard differential equation solver fails to actually solve the equation. That seems strange. Strange enough that it's worth sharing the error.

```
ODEintWarning: Excess work done on this call (perhaps wrong Dfun type).
  Run with full_output = 1 to get quantitative information.
  warnings.warn(warning_msg, ODEintWarning)
```

The error does not arise because of some error in coding; instead, it arises because of a fundamental change in the dynamics: the population blows up, explodes to infinity, encounters a singularity, or any one of similar euphemisms. What is striking is that, in contrast to the exponential growth model in which populations go to infinity—rapidly to be sure but in an infinite amount of time—the Malthus-Condorcet model reaches infinity in a finite time. Trying to identify that finite time underlies the next challenge.

CHALLENGE PROBLEM: Positive Feedback and Population Explosions

Modify and extend the following code to identify the finite time singularity in a logistic model with positive feedback, i.e., the Malthus-Condorcet model. In doing so, try to explain the choice to plot the time, $\hat{t}$ for $N(t)$, to cross a critical value $\hat{N}$ on inverse and log-scaled population axes. A working code should yield the following image:

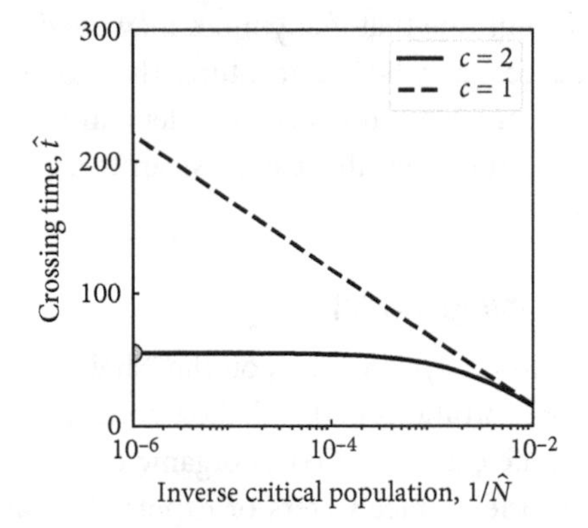

Notice that in the code below the actual dynamical system must also be coded up as a function. In addition, you will need to code up a modification to the standard numerical integration to check for crossing events. (Hint: For detecting crossings in Python, consider using the `np.argmax` function, which returns the first index that satisfies an input condition. Another option would be to use `scipy.integrate.solve_ivp` rather than the usual `scipy.integrate.odeint`. This new function would allow the user to write event detection code to terminate the calculation before any errors occur.)

```
# Crossing points c=2
Ncrit = np.logspace(2,6,1000)
t=np.arange(0,300,0.1)
pars['rho']=0.0001
pars['c']=2
pars['K0']=500
pars['N0']=50
y = integrate.odeint(...,args=(pars,),rtol=...)

# Find crossing points
tcrit = np.zeros(np.shape(Ncrit))
for i in range(len(Ncrit)):
    tmpi=np.argmax(...)
    tcrit[i]=...

plt.plot(...,linewidth=3)
```

13.4 SMALL MODELS OF A BIG CLIMATE

The final case study in this laboratory considers the problem of how and when a complex Earth system can change states. Given the spirit of the final chapter of the textbook, this last lab is a model of a model—but one that many more may need to understand if we are to avert the outcomes they suggest may lie in our future. The model was developed by Daniel Rothman (2017, 2019). It represents an effort to understand the mechanisms that led to mass extinctions in the past and what those events can teach us to prevent catastrophic extinctions in the present.

13.4.1 Baseline carbon cycle model

This model of the Earth's carbon cycle focuses on the "shallow" surface of the ocean, i.e., the photic zone. Carbon concentrations in this shallow zone reflect a combination of fluxes. Photosynthesis fixes inorganic carbon (CO_2) as organic carbon; this organic carbon may be remineralized back into the surface waters or exported from the surface to the deep waters (e.g., by aggregation, mortality, and sinking of living matter). Despite this complexity, the model focuses on the concentrations of dissolved inorganic carbon (predominantly CO_2). The rationale for this choice is that geochemical analyses of ancient sediments reveal changes in the ocean chemistry over time. These changes in ocean cheimstry serve as an indicator of the state of the oceans as well as a signature of extinctions. The main text provides an example of how variation in the isotopic composition of carbonate carbon reveals both the duration and magnitude of disruptions in the carbon cycle. The takeaway from the analysis of the geological record is that sufficiently large, sustained perturbations are strongly associated with mass extinctions. Yet, in the case of a rapid—i.e., singular—perturbation, it may be that sufficiently large perturbations (even when not sustained) can lead to massive disruption and mass extinction.

The model describes the dynamics of the carbonate ion concentration, $c(t) \equiv [CO_3^{2-}]$, and that of the total dissolved inorganic carbon (DIC), $w(t) \equiv [CO_2] + [HCO_3^-] + [CO_3^{2-}]$. Hence, when c is high, then the water has a high capacity to neutralize acids; i.e., this is the opposite in the case of ocean acidification. The model is as follows:

$$\dot{c}/f(c) = \mu\left[1 - bs(c, c_p) - \theta\bar{s}(c, c_x) - \nu\right] + w - w_0 \quad (13.11)$$

$$\dot{w} = \mu\left[1 - bs(c, c_p) + \theta\bar{s}(c, c_x) + \nu\right] - w + w_0 \quad (13.12)$$

given nondimensionalized time and $\mu = j_{in}\tau_w$ is a characteristic concentration. This carbon cycle model must be supplemented by a buffer function $f(c) = f_0 \frac{c^\beta}{c^\beta + c_f^\beta}$, a sigmoidal equilibrium of the carbonate system $s(c, c_p) = \frac{c^\gamma}{c^\gamma + c_p^\gamma}$, and $\bar{s} = 1 - s$. The full set of parameters is as follows: $\mu = 250$, $c_x = 70$, $c_p = 110$, $w_0 = 2000$, and $c_f = 43.9$, all in units of μmol/kg^{-1}, and the dimensionless parameters $b = 4$, $\theta = 8$, $\nu = 0$, $\gamma = 4$, $f_0 = 0.694$, and $\beta = 1.7$. For context, in the past 100 million years, the carbonate ion concentration has increased from $c \approx 50$ to $c > 200$; in addition, a baseline DIC level is approximately 2400 μmol/kg in the oceans. These values are useful for considering how deviations driven by anthropogenic input in the near term may shift the carbon cycle in the long term.

The following code represents the baseline model:

```
def ccycle(y,t,pars):
    # Carbon cycle model following Rothman PNAS (2019)

    # Variables
    c=y[0]
    w=y[1]

    # Parameters
    sc_cp = c**pars['gamma'] / (c**pars['gamma']+pars['cp']**pars['gamma'])
    sbar_c_cx = 1-c**pars['gamma']/(c**pars['gamma']+pars['cx']**pars['gamma'])
    fc =pars['f0']*(c**pars['beta']/(c**pars['beta']+pars['cf']**pars['beta']))

    # Model
    dydt = np.zeros(2)
    dydt[0] = fc*(pars['mu'] * \
      (1-pars['b']*sc_cp-pars['theta']*sbar_c_cx-pars['nu'])+w-pars['w0'])
    dydt[1] = pars['mu'] * \
      (1-pars['b']*sc_cp+pars['theta']*sbar_c_cx+pars['nu']) - w + pars['w0']
    return dydt
```

CHALLENGE PROBLEM: Simulating Carbon (Limit) Cycles

Simulate the model using two distinct starting conditions to demonstrate that this model can yield an attracting limit cycle, even in the absence of sustained CO_2 injection, $\nu = 0$. The following phase plane images illustrate what such dynamics should look like given two initial conditions: (i) $c = 25, w = 2600$, and (ii) $c = 200, w = 3250$, in the cases $\theta = 2$ (left) and $\theta = 6$ (right). As is evident, the same model can lead to convergence to equilibrium as well as stable limit cycles.

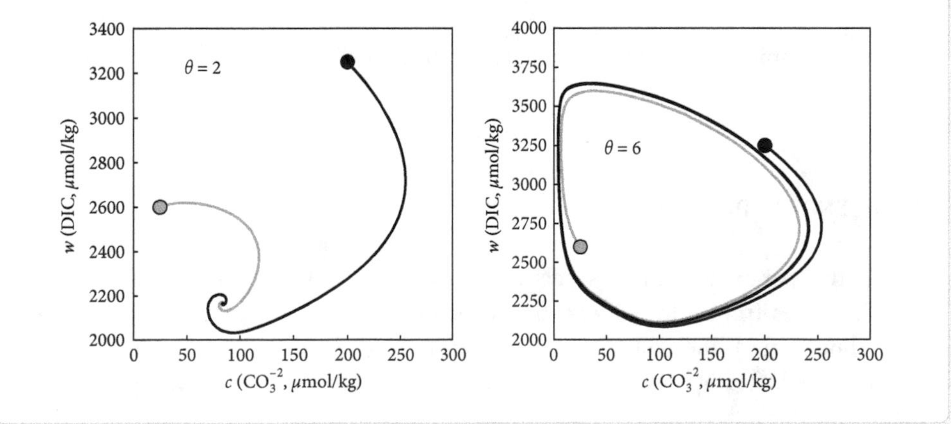

13.4.2 Transition to excitability

The carbon cycle model can also exhibit excitability, i.e., show different responses to small and large perturbations. The relevance of this is particularly apt in the case of injections of CO_2 over short time scales, e.g., via vulcanism or anthropogenic emissions. The consequences, however, are radically different than many conventional perspectives. The intrinsic dynamics of the carbon cycle make it possible that a long-term excitatory cascade can unfold given relatively small differences in the injection. Below is an example of how changes in ν lead to increasingly large excitations despite the fact that the steady state values remain relatively unchanged. Using the same parameters as before, except changing $c_x = 55$, you should try to recapitulate these dynamics:

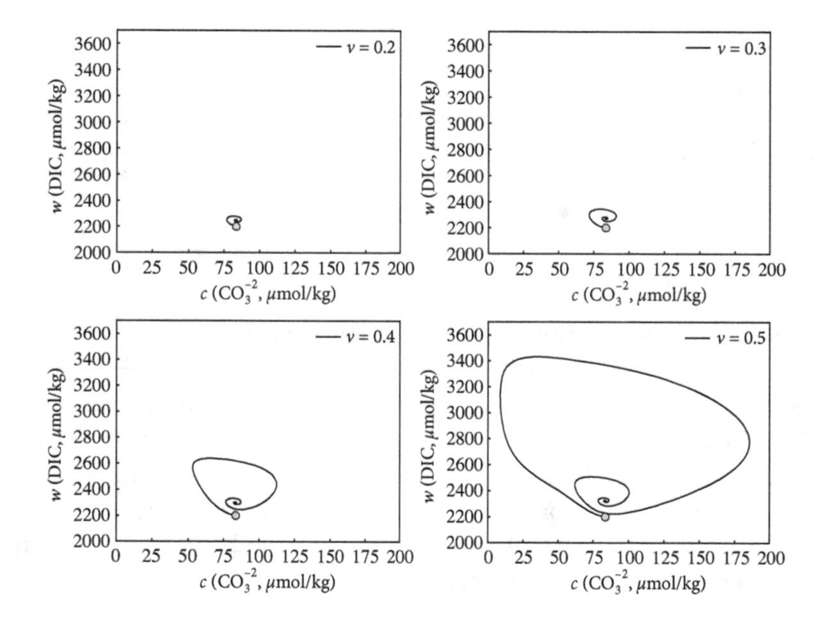

These dynamics also relate to the final challenge problem.

CHALLENGE PROBLEM: Excitations in the Carbon Cycle

Using the parameters in this section, contrast the changes in the fixed point as ν increases from 0 to 0.5 with the scope of the excitations. If your code works, you should find the following result, including both the excursion size and the expected steady state. Compare and contrast the curves and interpret this finding.

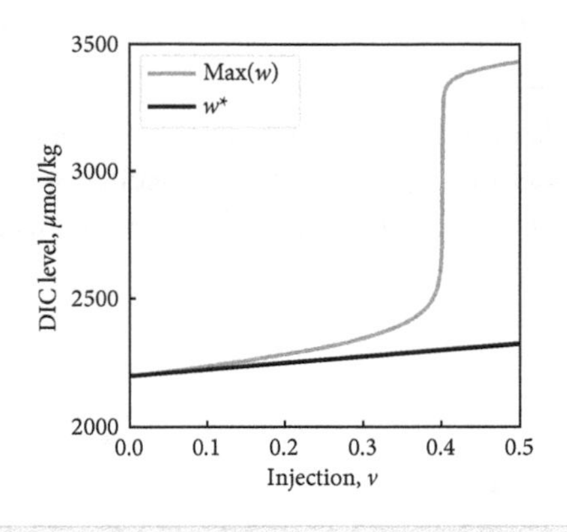

13.5 CODA

The work is done. You did it. Yet there is more to do, whether in controlling the dynamics of single cells and the spread of infectious disease, developing evolutionarily robust approaches to treat antibiotic-resistant pathogens, improving understanding of the brain (both when it is healthy and when it is not), or characterizing the fate of interwoven ecosystems. These computational laboratories are intended to provide the kind of tacit knowledge needed to transform the study of living systems into one infused with mathematical theory, physical intuition, and computational models. Your skills will get better with practice, repetition, exploration, and perhaps even with some risk taking of your own to develop new models where none yet exist.

SOLUTIONS TO CHALLENGE PROBLEMS

SOLUTION: Finding the Equilibria

The equilibrium point for the dynamics satisfies $N(t+1) = N(t) = N^*$. This is true for $N^* = 0$ and for $N^* = 1 - R^{-1}$. Hence, the predicted dynamics should be 1/6, 1/3, and 4/9 for the values of $R = 1.2$, 1.5, and 1.8, respectively.

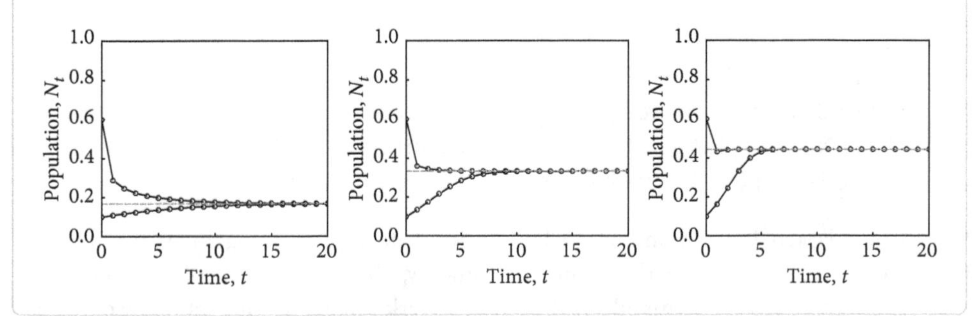

SOLUTION: Cobwebbing Dynamics

One approach to this problem is to break this into three components. First, plot the map. Second, overlay the one-to-one line. Third, simulate the model step by step and plot a sequence of two-line segments (as noted in the hint). The solution also includes labeling commands. The answers are as follows:

```
# Parameters
pars={}
pars['Rvals']=[2.8, 3.2, 3.9]

rNum=len(pars['Rvals'])
fig, axes = plt.subplots(1,3,figsize=(5*rNum+2*(rNum-1),5))
for i in range(rNum):
    ax = axes[i]

    # Conditions
    pars['R']=pars['Rvals'][i]
    pars['tmax']=25
    x0=0.1
    # The model
    disc_logistic = lambda x: pars['R']*x*(1-x)

    # The map
    xrange =np.arange(0,1.001,0.001)
    ax.plot(xrange,disc_logistic(xrange),'k',linewidth=3)
    ax.plot(xrange,xrange,'k',linewidth=2,linestyle='--')

    # Simulate
    x = np.zeros(pars['tmax']+1)
    x[0]=x0
    for n in range(pars['tmax']):
        x[n+1]=disc_logistic(x[n])
        ax.plot([x[n], x[n], x[n+1]],[x[n], x[n+1], x[n+1]],
             c=[0.5,0.5,0.5], linewidth=2)
    # Annotate
    ax.set_ylim(0,1)
    ax.set_xlim(0,1)
    ax.set_xlabel(r'Current state, $x_t$',fontsize=20)
    ax.set_ylabel(r'Next state, $x_{t+1}$',fontsize=20)
    ax.set_title(r'$R={R}$'.format(R=pars['R']),fontsize=20)
```

Note that this code can be adapted into an animation by pausing before plotting again in the figure. Give the command `%matplotlib qt` in the command window, then at the end of the loop say `plt.pause(0.01)` and `plt.show()`. This allows the code to overlay new lines at a rate slow enough to see. Alternatively, use the `matplotlib.animation` package. The code in the loop will be encoded in an update function. The resulting animation will come complete with a play button. This is an exercise for the very interested reader.

SOLUTION: Continuous Logistic Model from Scratch

The following code represents one way to solve this problem, but not the only way. Here the solution separates the function from the script for later use and reuse. First the script:

```
# Model specification
t = np.arange(0,200,0.1)
pars={}
pars['r']=0.1
pars['K']=10**5
N0=50

# Simulate
y= integrate.odeint(logistic_model,N0,t,args=(pars,))

# Plot and label
fig = plt.figure(figsize=(5,5))
ax = fig.gca()
plt.plot(t,y,'k',linewidth=3)
plt.xlabel('Time',fontsize=20)
plt.ylabel('Population',fontsize=20)
plt.xlim(min(t),max(t))
plt.ylim(0,1.1*pars['K'])
```

Then the function:

```
def logistic_model(N,t,pars):
    # Logistic model dynamics
    # dNdt = rN*(1-N/K)
    dNdt = pars['r']*N*(1-N/pars['K'])
    return dNdt
```

This code yields the diagram shown on the left. And adding `scale_y_continuous(trans='log10')` to our plotting commands can also yield the figure on the right, where the initial exponential growth is self-evident given the straight-line dynamics on a log-scaled *y* axis.

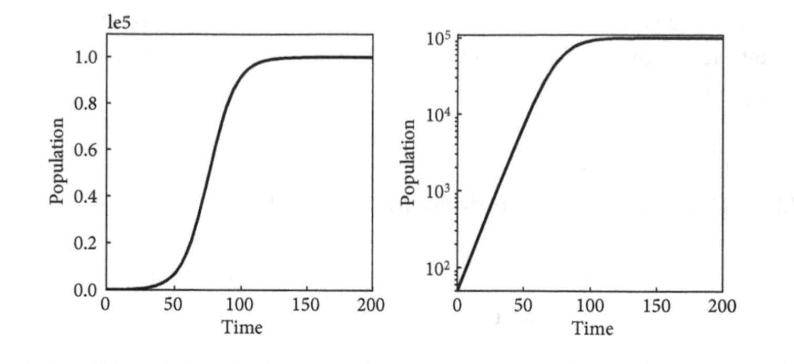

SOLUTION: Positive Feedback and Population Explosions

The following code includes a script that simulates the model with fixed parameters and systematically finds the time in which the population, $N(t)$, crosses a series of thresholds. The code then plots $1/\hat{N}$—the thresholds—on the x axis against the crossing time, $\hat{t}$, on the y axis. Note that for exponential growth $\hat{t} \sim log(\hat{N})$, so that there is no finite time convergence to an infinite population for $c=1$ in the Malthus-Condorcet model. The solution has one additional part, a crossing code, also included here. First, the main code:

```
# Crossing points c=2
Ncrit2 = np.logspace(2,5.5,1000)
t=np.arange(0,200,0.0001)
pars['rho']=0.0001
pars['c']=2
pars['K0']=500
pars['N0']=50
y = integrate.odeint(condorcet_model,np.array([pars['N0'],pars['K0']]),t,
                     args=(pars,),rtol=10**-8)

# Find crossing points c=2
tcrit2 = np.zeros(np.shape(Ncrit2))
for i in range(len(Ncrit2)):
    tmpi=np.argmax(y[:,0]>Ncrit2[i]) # Find first point that exceeds value
    tcrit2[i]=t[tmpi]

# Plot
fig = plt.figure(figsize=(5,5))
ax = fig.gca()
plt.plot(1/Ncrit2,tcrit2,'k',linewidth=3)
ax.set_xscale('log')

# Crossing points c=1
Ncrit = np.logspace(2,6,1000)
t=np.arange(0,400,0.0001)
```

```
pars['rho']=0.0001
pars['c']=1
pars['K0']=500
pars['N0']=50
y = integrate.odeint(condorcet_model, np.array([pars['N0'],pars['K0']]),t,
                     args=(pars,),rtol=10**-8)

# Find crossing points
tcrit = np.zeros(np.shape(Ncrit))
for i in range(len(Ncrit)):
    tmpi=np.argmax(y[:,0]>Ncrit[i])
    tcrit[i]=t[tmpi]

plt.plot(1/Ncrit,tcrit,'k',linestyle='--',linewidth=3)
```

Additional plotting commands

Here, as a bonus, are the plot labeling commands:

```
# Singular point
plt.plot([10**-6, np.min(1/Ncrit2)],[54.9306, np.max(tcrit2)],'k',linewidth=3)
plt.plot(10**-6, 54.9306,marker='o',
         markeredgecolor='k',markerfacecolor=[0.75,0.75,0.75],markersize=12)

plt.legend([r'$c=2$', r'$c=1$'],fontsize=15,loc='upper right')
plt.xlabel(r'Inverse critical population, $1/\hat{N}$',fontsize=20)
plt.ylabel('Crossing time, $\hat{t}$',fontsize=20)
plt.setp(ax.spines.values(),linewidth=2)
ax.tick_params(labelsize=15,direction='in',width=2)
plt.xticks([10**-6,10**-4,10**-2])
plt.yticks([0,100,200,300])
```

SOLUTION: Simulating Carbon (Limit) Cycles

The solution here arises from a standard case of entering in parameter values and simulating a dynamical system. The following is a baseline code suitable for modification:

```
# Main data goes here - units in Table S1 of Rothman
# umol/kg/yr

pars={}
pars['mu']=250
pars['cx']=70
```

```
pars['cp']=110
pars['w0']=2000
pars['cf']=43.9

# Dimensionless
pars['b']=4
pars['theta']=2 # Modify this value
pars['nu']=0
pars['gamma']=4
pars['f0']=0.694
pars['beta']=1.70
pars['jin']=0.025
pars['tauw']=10**4

# Simulation
t=np.arange(0,100,0.001)
pars['y0']=np.array([25, 2600])
y = integrate.odeint(ccycle, pars['y0'],t,args=(pars,),rtol=10**-8)
plt.plot(y[:,0],y[:,1],c=[0.5,0.5,0.5],linewidth=3)
plt.plot(y[0,0],y[0,1],
         marker='o',
         markersize=12,
         markerfacecolor=[0.5,0.5,0.5],
         markeredgecolor='k')

pars['y0']=np.array([200, 3250])
y = integrate.odeint(ccycle,pars['y0'],t,args=(pars,),rtol=10**-8)
plt.plot(y[:,0],y[:,1],'k',linewidth=3)
plt.plot(y[0,0],y[0,1],
         marker='o',
         markersize=12,
         markerfacecolor='k',
         markeredgecolor='k')
```

SOLUTION: Excitations in the Carbon Cycle

Using the same parameters as above, the following code presents the change in steady state (which is relatively modest) with that maximum excursion given an initial condition of $w = 2000$ μmol/kg. The key takeaway is that, although the effects of increases in ν seem modest (in terms of equilibrium), there is a dramatic change in the magnitude of the excitation, very similar to that seen in neuronal responses close to the value of $\nu = 0.4$.

```
# Steady states
cstar = (pars['b']-1)**(-1/pars['gamma'])*pars['cp']
wstar = pars['w0']+pars['mu']*(pars['theta']+pars['nu']-pars['theta']*\
          pars['cp']**pars['gamma']/((pars['b']-1)*\
              pars['cx']**pars['gamma']+pars['cp']**pars['gamma']))

# Perturbations
pars['y0']=np.array([cstar, 2200])
pars['nurange']=np.arange(0,0.501,0.001)

t=np.arange(0,100,0.001)
wmax = np.zeros(np.shape(pars['nurange']))
ystar = np.zeros((len(pars['nurange']),2))
wstar_nu = np.zeros(np.shape(pars['nurange']))
for i in range(len(pars['nurange'])):
    # Simulation compare to nu short
    pars['nu']=pars['nurange'][i]
    cstar = (pars['b']-1)**(-1/pars['gamma'])*pars['cp']
    wstar = pars['w0']+pars['mu']*(pars['theta']+pars['nu']-pars['theta']*\
          pars['cp']**pars['gamma']/((pars['b']-1)*\
              pars['cx']**pars['gamma']+pars['cp']**pars['gamma']))
    y = integrate.odeint(ccycle,pars['y0'],t,args=(pars,),rtol=10**-8)
    #wstar = y[-1,1]
    wmax[i] = np.max(y[:,1])
    ystar[i,:] = np.array([cstar,wstar])
    wstar_nu[i] = wstar

fig = plt.figure(figsize=(5,5))
ax = fig.gca()
plt.plot(pars['nurange'],wmax,c=[0.5,0.5,0.5],linewidth=3)
plt.plot(pars['nurange'],wstar_nu,'k',linewidth=3)

plt.xlabel(r'Injection, $\nu$',fontsize=20)
plt.ylabel(r'$DIC$ levels, $\mu mol/kg$',fontsize=20)
```

BIBLIOGRAPHY

Cohen, J. E. (1995). Population growth and earth's human carrying capacity. *Science*, 269(5222):341–346.

Fisher, R. A. (1958). *The Genetical Theory of Natural Selection*. Dover, New York, 2nd edition.

FitzHugh, R. (1961). Impulses and physiological states in theoretical models of nerve membrane. *Biophysical Journal*, 1(6):445–466.

Gardner, T. S., Cantor, C. R., and Collins, J. J. (2000). Construction of a genetic toggle switch in *Escherichia coli*. *Nature*, 403:339–342.

Gillespie, D. T. (1977). Exact stochastic simulation of coupled chemical reactions. *Journal of Physical Chemistry*, 82(25):2340–2361.

Hodgkin, A. L., and Huxley, A. F. (1952). A quantitative description of membrane current and its application to conduction and excitation in nerve. *Journal of Physiology*, 117:500–544.

Hofbauer, J., and Sigmund, K. (1998). *Evolutionary Games and Population Dynamics*. Cambridge University Press, Cambridge, UK.

Izhikevich, E. M. (2007). *Dynamical Systems in Neuroscience*. MIT Press, Cambridge, MA.

Kaiser, D. (2005). *Drawing Theories Apart: The Dispersion of Feynman Diagrams in Postwar Physics*. University of Chicago Press, Chicago.

Keeling, M. J., and Rohani, P. (2007). *Modelling Infectious Diseases*. Princeton University Press, Princeton, NJ.

Lotka, A. (1925). *Elements of Physical Biology*. Dover, New York, reprinted 1956 edition.

Luria, S. E., and Delbrück, M. (1943). Mutations of bacteria from virus sensitivity to virus resistance. *Genetics*, 28:491–511.

May, R. M. (1979). Simple mathematical models with complicated dynamics. *Nature*, 261(10):3.

McNally, L., Bernardy, E., Thomas, J., Kalziqi, A., Pentz, J., Brown, S. P., Hammer, B. K., Yunker, P. J., and Ratcliff, W. C. (2017). Killing by type VI secretion drives genetic phase separation and correlates with increased cooperation. *Nature Communications*, 8(1):14371.

Moran, P. A. P. (1958). Random processes in genetics. *Proceedings of the Cambridge Philosophical Society*, 54:60–71.

Nagumo, J., Arimoto, S., and Yoshizawa, S. (1962). An active pulse transmission line simulating nerve axon. *Proceedings of the IRE*, 50(10):2061–2070.

Nowak, M. A. (2006). *Evolutionary Dynamics: Exploring the Equations of Life*. Belknap Press of Harvard University Press, Cambridge, MA.

Press, W. H., Vetterling, W. T., Teukolsky, S. A., and Flannery, B. P. (1986). *Numerical Recipes*, volume 818. Cambridge University Press, Cambridge, UK.

Purcell, E. M. (1977). Life at low Reynolds number. *American Journal of Physics*, 45(1):3–11.

Rosenblueth, A., and Wiener, N. (1945). The role of models in science. *Philosophy of Science*, 12:316–321.

Rosenzweig, M. L., and MacArthur, R. H. (1963). Graphical representation and stability conditions of predator-prey interactions. *American Naturalist*, 97:209–223.

Rothman, D. H. (2017). Thresholds of catastrophe in the Earth system. *Science Advances*, 3(9):e1700906.

Rothman, D. H. (2019). Characteristic disruptions of an excitable carbon cycle. *Proceedings of the National Academy of Sciences*, 116(30):14813–14822.

Shapere, A., and Wilczek, F. (1989). Geometry of self-propulsion at low Reynolds number. *Journal of Fluid Mechanics*, 198:557–585.

Strogatz, S. (1994). *Nonlinear Dynamics and Chaos*. Addison Wesley, Reading, MA.

Tritton, D. P. (1988). *Physical Fluid Dynamics*. Oxford University Press, Oxford, UK.

Vicsek, T., Czirók, A., Ben-Jacob, E., Cohen, I., and Schochet, O. (1995). Novel type of phase transition in a system of self-driven particles. *Physical Review Letters*, 75:1226–1229.

Volterra, V. (1926). Fluctuations in the abundance of a species considered mathematically. *Nature*, 118:558–60.

Wakeley, J. (2008). *Coalescent Theory: An Introduction*. Roberts and Company, Greenwood Village, CO.

Wright, S. (1932). The roles of mutation, inbreeding, crossbreeding and selection in evolution. *Proceedings of the 6th International Congress of Genetics*, 1:356–366.

Printed in the USA
CPSIA information can be obtained
at www.ICGtesting.com
JSHW052019130524
63048JS00008B/387

9 780691 255675